Patterns of human growth

Cambridge Studies in Biological Anthropology

Series Editors

G. W. Lasker
Department of Anatomy, Wayne State University, Detroit, Michigan, USA

C. G. N. Mascie-Taylor
Department of Physical Anthropology, University of Cambridge, UK

D. F. Roberts
Department of Human Genetics, University of Newcastle-upon-Tyne, UK

Patterns of human growth

BARRY BOGIN

Associate Professor of Anthropology,
Department of Behavioral Sciences,
University of Michigan–Dearborn, USA

The right of the
University of Cambridge
to print and sell
all manner of books
was granted by
Henry VIII in 1534.
The University has printed
and published continuously
since 1584.

CAMBRIDGE UNIVERSITY PRESS

Cambridge
New York   New Rochelle
Melbourne   Sydney

Published by the Press Syndicate of the University of Cambridge
The Pitt Building, Trumpington Street, Cambridge CB2 1RP
32 East 57th Street, New York, NY 10022, USA
10 Stamford Road, Oakleigh, Melbourne 3166, Australia

First published 1988

Printed in Great Britain at the University Press, Cambridge

British Library cataloguing in publication data
Bogin, Barry
  Patterns of human growth. – (Cambridge studies in
  biological anthropology; 3).
  1. Human growth
  I. Title
  612.6   QP84

Library of Congress cataloguing in publication data
Bogin, Barry.
  Patterns of human growth.
  (Cambridge studies in biological anthropology; 3)
  Bibliography: p.
  Includes index.
  1. Human growth.   2. Physical anthropology.
  3. Human evolution.   I. Title.   II. Series.
  GN62.9.B64   1988   573   87-23913

ISBN 0 521 34593 6 hard covers
ISBN 0 521 34690 8 paperback

Dedicated to Gabriel Ward Lasker in admiration and friendship and Sandra Lynn Bogin with love

Contents

Acknowledgments

The author is very fortunate to have several colleagues who took time from their own research to read and criticize one or more chapters of this book. For their kindness, and for their forthright comments which have improved the presentation of this work, sincere thanks are extended to Drs George Clark, Nancy Howell, Marquisa LaVelle, Michael A. Little, Daniel Moerman, Gerald Moran, B. Holly Smith, Jessica Schwartz, and Elizabeth Watts.

Part of this book was written during a research leave from the University of Michigan–Dearborn. The author appreciates the assistance of Drs Eugene Arden, Victor Wong, and Donald Levin in helping to arrange for the leave. A generous grant from the American Philosophical Society helped to defray research expenses.

Sandra Bogin contributed several of the line drawings that illustrate the text and, more importantly, provided encouragement for the writing. Prof. Gabriel W. Lasker read every word of the draft of the text, and engaged the author in many discussions about the intellectual and technical content of the book. His involvement in all aspects of the production of the book are appreciated deeply.

Introduction

It is the purpose of this book to describe and interpret some of the evolutionary, physiological, cultural, and mathematical patterns of human growth. Given this purpose, the title of this book requires some explanation. A cell biologist might think of the phrase 'patterns of growth' in terms of a series of DNA controlled cell duplication and division events. An embryologist might think of patterns of cell differentiation and integration leading to the development of a functionally complete human. The clinician interprets patterns of growth, especially deviations from expected or 'normal' growth, as evidence of disease or other pathology in the patient. All of these concepts of 'pattern' may be biologically valid and useful in their own areas of specialization, but this book is about none of them. The goal of this account is to consider the growth of the human body in a unified and holistic manner. The result, it is hoped, will be a synthesis of the forces that shaped the evolution of the human growth pattern, the biocultural factors that direct its expression in populations of living peoples, the intrinsic and extrinsic factors that regulate individual development, and the biomathematical approaches needed to analyze and interpret human growth.

Growth and evolution

Dobzhansky (1973) said that 'nothing in biology makes sense except in the light of evolution.' Human growth, which follows a unique pattern among the mammals and, even, the primates, is no exception to Dobzhansky's admonition. A consideration of the chimpanzee and the human, perhaps the two most closely related (genetically) extant primates, shows the value of taking an evolutionary perspective on growth. Huxley (1863) demonstrated many anatomical similarities between chimpanzees and humans. King & Wilson (1975) showed that such anatomical similarities are due to a near identity of the structural DNA of the two species. One interpretation of King & Wilson's findings is that the differences in size and shape between chimpanzee and human are due to the regulation of gene expression, rather than the possession of unique genotypes. Of course, humans are not descended from the chimpanzee, but both species did have a common ancestor some six, or more, million

1

years ago. During evolutionary time, mutations and selective forces were at work on the descendants of this ancestor shaping their genetic constitution and its expression in their phenotypes.

The anatomical differences between human and chimpanzee that result from alterations in gene regulation are achieved, in part, through the alterations in growth rates. D'Arcy Thompson (1917) was one of the first scholars to show that the differences in form between the adults of various species may be accounted for by differences in growth rates from an initially identical – one might better say 'similar' – form. Thompson's transformation grids (Figure I.1) of the growth of the chimpanzee and human skull from birth to maturity, showed how both may be derived from a common neonatal form. Different patterns of growth of the cranial bones, maxilla, and mandible are all that are required to produce the adult differences in skull shape. The same holds true for the post-cranial skeleton.

Figure I.1. Transformation grids for the chimpanzee and human skull during growth. Fetal skull proportions are shown above for each species. The relative amount of distortion of the grid lines overlying the adult skull proportions indicates the amount of growth of different parts of the skull (after D'Arcy Thompson, 1942).

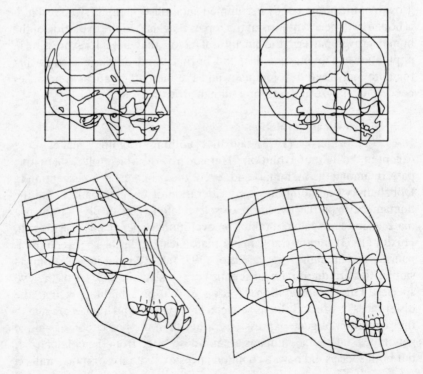

Despite the anatomical and biochemical evidence for the evolutionary origins of the human growth pattern, most works on human growth give little consideration to this topic. A paragraph or two is all that may be found in the current anthropological, physiological, and medical texts devoted to human growth. In the book before you, however, two chapters are devoted to an account of the evolution of the human pattern of growth. These two chapters follow an overview of the history of the study of growth and the basic biological principles of growth and development during the human life cycle, presented in Chapter 1. Chapter 2 describes the evolution of the human growth curve. The mammalian foundations of the human growth curve, and the non-human primate embellishments upon those foundations, are presented conceptually and mathematically. The unique features of the human pattern of growth, including the evolution of the childhood growth period and the human adolescent growth spurt in height, are detailed. In Chapter 3 the pattern of human growth is considered from an ecological and evolutionary perspective. The relation of growth rates to feeding and reproductive adaptations is explored using data from paleontology, paleoecology, demography, ethology, and ethnology. The result is a comprehensive explanation for the functional and adaptive significance of human growth patterns.

Growth theory

This is also a book about a theoretical approach to the study of human growth. The literature in the general area of animal growth is rich in both hypothesis testing and theory (e.g. Huxley, 1932; Thompson, 1942; Weiss & Kavanau, 1957; Bertalanffy, 1960; Goss, 1964; 1978). By contrast, only a few others have published on theories about the course and regulation of human growth (e.g. Tanner, 1963; Frisch & Revelle, 1970; Grumbach *et al.*, 1974; Bogin, 1980).

Most research and writing in the area of human growth and development is descriptive and non-theoretical. Typically, in fields such as anthropology and epidemiology, a sample of children or adults is measured for some variable(s) (height, weight, etc.) and the data are presented, described, and compared with similar data for another group of people. Or, in the fields of medicine and psychology, an unusual case study of growth, perhaps resulting from physical or psychosocial pathology, is described. In another realm, growth data are derived from statistically representative samples of human populations and are used to construct reference standards of height, weight, and other physical dimensions. These standards have value in public health work to assess the growth, development, and nutritional status of populations that are 'at risk' for growth failure or malnutrition, and to monitor the effectiveness of intervention programs designed to improve the health and growth status of such populations.

Historically, most disciplines begin with this sort of descriptive phase. Human growth research is no exception, as demonstrated in two recently published books that detail the historical development of the field (Boyd, 1981; Tanner, 1981). One sign of a maturing discipline is when it begins to develop hypotheses to examine the nature of the processes that account for the descriptive data. An example of the use of hypotheses in growth research is the classic series of studies by Boas (e.g. 1912; 1922) on the growth of the children of migrants to the United States. Boas challenged the dogmatic belief that physical types (i.e., races) were fixed and biologically distinct. His hypothesis was that a change in environment, including changes in nutrition and child rearing practices, would bring about alterations in the amount and rate of growth. By studying adults he showed how such changes in growth had modified racial hallmarks, for example, the cephalic index (a measure of skull shape), to an extent that would obliterate the distinctions between the supposed racial types. Boas was right, and the days of cephalometric and, generally, anthropometric racial classification are long past. The value of his hypothesis testing method lasted, and it is still being used in the area of migration research to test the relationship between migration history and human biology (Boyce, 1984; Mascie-Taylor & Lasker, 1987).

Mature sciences are noted for their ability to synthesize several hypotheses that have been independently verified into comprehensive theories that can explain the known data and, in turn, indicate the kind of observations that should be made in further research. Human growth research is now entering the hypothesis testing stage. The relationships between nutrition and growth, physical development and chronic disease, and environmental stress and growth, among other topics, are being actively pursued although, unfortunately, a dreadful number of purely descriptive studies are still being produced.

Human and non-human animal growth research

Training may explain some of the emphasis on description in human growth research and the comparative wealth of hypothesis testing in growth research with non-human animals. Human biologists are often trained in departments of anthropology where the emphasis is on fieldwork and field research methodologies. Others are trained in the 'health sciences,' where the emphasis is on clinical or epidemiological applications. In the field, or the clinic, human growth must be viewed as a process that takes place in a social and cultural setting. The Pygmalion question of the relative contribution of genetics and environment to human development is a consequence of the biocultural nature of humans. A specific example is the growth status of Guatemalan Indian children. These children are, on average, shorter and lighter than non-Indian, that is, 'westernized,' children in Guatemala. Indians are even smaller than the poorest non-Indian children living in slums and squatter settlements in Guatemala City (Bogin & MacVean, 1984; Johnston *et al.*, 1985). The possible genetic contribution to Indian short stature is very difficult to determine, however, due to the fact that virtually all Indians are of lower socioeconomic status than non-Indians. The sequelae of low socioeconomic status, including undernutrition and infectious disease, confound and probably override any genetic difference between the two ethnic groups.

Researchers studying the growth of non-human animals are more often trained in the biological laboratory, and are exposed to animal growth as a problem defined in physical and biochemical terms. In the laboratory, genetic and environmental determinants of growth may be studied with precision. The growth of anatomical regions, even isolated populations of cells, of the body may be controlled and, therefore, understood. It is almost always impossible to justify on ethical grounds conditions which impose this level of experimental control in the study of human growth. It is also conceptually unrealistic, since normal human growth and development require the complexities of a normal social and cultural milieu. Even so, the human biologist is not restricted to descriptive studies, leaving the general biologist to pursue the more intellectually exciting experimental and theoretical research. A combination of creative and rigorous field research, powerful new statistical and computational methods, study of pathological growth disorders, and experimentation within allowable conditions can be used to achieve a better understanding of human growth.

Human auxology

This book is a synthesis of methods and knowledge about human growth gleaned from evolutionary biology, from reports of the growth of human populations living under various ecological regimes, from statistical and mathematical applications, from medical pathology, and from experimental biology. Drawing upon these areas of research, the book strives to include human growth within the field of theoretical auxology. The term auxology refers to the study of biological growth. It could be the study of any type of growth, and some botanists and most veterinary and farm animal zoologists use this term to refer to growth research in their fields. During the last decade, European human biologists began to use the phrase 'human auxology' to refer to human growth research (Borms *et al.*, 1984; Bogin, 1986). Theoretical auxology combines descriptive studies, experimental research, and hypothesis testing into a comprehensive view of the structural and functional elements of growth.

The contribution toward theoretical auxology made in this book attempts to show that the tempo and mode of human growth are basic to the understanding of our species' place in nature. This is done by dividing the patterns of human growth into four areas, each area treated in separate chapters. The first area is the evolutionary foundation of the human growth pattern, treated in Chapter 2 and Chapter 3 as described previously. The second area is population variation in growth patterns. In Chapter 4, several cases of such variation are described and the adaptive value of population differences in growth is discussed from an evolutionary perspective. In Chapter 5, some of the physiological, environmental, and cultural reasons for population variation in growth are explored using the literatures of field and experimental research on human growth.

The third theoretical area is covered in Chapter 6, which describes and analyzes the genetic and endocrine factors that regulate the growth of individual human beings. The facts, explanations, and hypotheses presented in the previous chapters give evolutionary and functional meaning to the pattern of human growth, but they do not explain how the amount and rate of growth are controlled. Genes provide the structural elements (e.g., proteins) for growth and form. Genetic information also interacts with factors from the environment, and these factors help to provide guidelines for genetic expression. The gene–environment interactions may not, however, directly regulate growth and development. Rather, their influence is often mediated by the endocrine system. In Chapter 6, several examples of hormonal mediation are given, including the control of small body size in African pygmies, and the effects of nutritional and psychological stress on growth.

Chapter 7 is devoted to the fourth theoretical area, mathematical and biological models of the process of human growth and development. Some of the classic and recent innovative quantitative approaches to the study of growth regulation are presented. Qualitative approaches to the study of growth regulation and control are reviewed. A new model of growth regulation is presented that combines quantitative and qualitative elements from mathematics, molecular biology, and neuroendocrine physiology. As a challenge for the future, this chapter ends with a proposal for new areas of biological and mathematical research that may be of some value.

1   *Basic principles of human growth*

People, like all animals, begin life as a single cell, the fertilized ovum. Guided by the interaction of the genetic information provided by each parent and the environmental milieu, this cell divides and grows, differentiates and develops into the embryo, fetus, child, and adult. Though growth and development may occur simultaneously, they are distinct biological processes. *Growth* may be defined as a quantitative increase in size or mass. Measurements of height in centimeters or weight in kilograms indicate how much growth has taken place in a child. Additionally, the growth of a body organ, such as the liver or the brain, may also be described by measuring the number, weight, or size of cells present. *Development* is defined as a progression of changes, either quantitative or qualitative, that lead from an undifferentiated or immature state to a highly organized, specialized, and mature state. Maturity, in this definition, is measured by functional capacity, for example, the development of motor skills of a child as these skills are related to maturation of the skeletal and muscle systems. Though broad, this definition allows one to consider the development of organs (e.g., the kidney), systems (e.g., the reproductive system), and the person.

Historical background for the study of human growth*

To understand what is currently known about human growth and development it is useful to review their history of study. This history is important from a practical viewpoint, since it shows what people have studied, when such inquiry first occurred, and problems that are in need of further study. The history is important from a conceptual perspective as well, since it may help explain why scholars and practitioners have been interested in human development. Archaeological records from ancient Mesopotamian civilizations show that the earliest interest in the biology of children was primarily a concern with the preservation of life. Greek, Roman, and Arab physicians prescribed regimes of physical activity, education, and diet to help assure the health of children, but their advice was guided by the needs of their societies for military personnel and

* Information for this section was compiled from Lowery, 1986; Boyd, 1981; Tanner, 1981. Greater historical detail is given in each of these sources.

7

religious dogma about children, not by empirical observations of the effect of child rearing practices on child growth, development, and health. It was not until the 16th century that a scientific approach was taken by scholars and physicians to the study of child growth. The following is a brief review of some of the major historical events in the study of human growth, with special emphasis on those which still have an influence on growth research today.

The process of growth and development from fertilized ovum to the birth of a human child is so counterintuitive to our expectations, based on our experience with child growth after birth, that through much of human history scholars and physicians did not know or believe that it occurred. It was not until the year 1651 that the physician William Harvey helped establish that the embryo is not a preformed adult, rather that during development the human being passes through a series of embryological stages that are distinct in appearance from the form visible just before and after birth. The Greek physician Galen (*c.* AD 130 to *c.* AD 200) wrote about the appearance of the fetus in the later stages of pregnancy. However, the first accurate drawings of the fetus were made by Leonardo da Vinci (1452–1519), who dissected a seven month old fetus and stillborn, full-term infants. Other descriptions of fetal anatomy and physiology followed Leonardo's work, notably the studies published by Vesalius in 1555 and Volcher Coiter in 1572. During the 17th and 18th centuries descriptive anatomical studies continued, with most of the work being done on fetuses seven months old or older. The fetus of this age is of unmistakable human appearance, so these studies failed to appreciate the physical changes that take place earlier in prenatal life. Some biologists continued to believe in preformation, and a few extended that concept beyond pregnancy to the formation of spermatozoa (Figure 1.1). In 1799 S. T. Sommerring published drawings of the human embryo and fetus from the fourth week after fertilization to the fifth month. These drawings clearly showed that the embryo is not a preformed, or miniature, human being.

The scientific study of the cellular mechanisms of fertilization and embryonic development has its roots in the work of Karl Von Baer, published in 1829. Baer described the 'germ layers' of the embryo, properly called the endoderm, the mesoderm, and the ectoderm. The endoderm cells give rise to the internal organs, cells from the mesoderm form the skeleton and the muscles, and the skin and the teeth develop from ectoderm cells. Baer's work removed the need to invoke mystical 'vital forces' to explain embryological transformations, replacing these with more mechanistic processes. However, it was not until the 20th century that an understanding was achieved of the highly complex nature

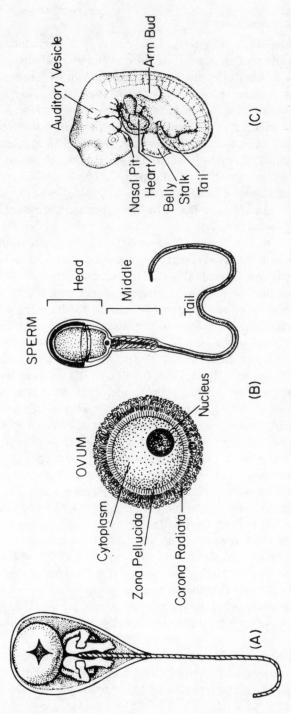

Figure 1.1. (A) Preformationist rendering of a human spermatozoon (after Hartsoeker, 1694; from Singer, 1959). (B) Diagram of human ovum and sperm. (C) Diagram of human embryo 32 days after fertilization.

of the physical, chemical, and biological processes that occur during prenatal growth.

The study of growth after birth, like the study of prenatal growth, began with a concept of infants and children as miniature adults who only had to increase in size during the growing years. Prior to the time of the Renaissance in Europe, physicians followed the Greek, Roman, and Arab traditions of treating the child as a miniature adult. Pre-Renaissance artists depicted children in the same way; the proportional differences between children and adults, for instance, the relatively larger head and shorter extremities of children when compared with adults, were not correctly rendered in paintings of this period. The *Rucellai Madonna*, attributed to the Italian artist Duccio (1285?), the *Madonna of the Trees* by Bellini (1487), and *Peasant Dance* by Pieter Bruegel the elder (1568?) are all in this stylistic tradition.

By the late 15th century the representation of children in art was changing. Leonardo's drawings correctly rendered adult and child body proportions. Albrecht Dürer (1471–1528) devised a method of geometric transformations that he used to accurately render proportions of the human head and face. He applied his method to drawings of men, women, and children. Including women and children in this type of methodological work was an innovation, since most artists followed the teachings of Cennino Cennini (*c.* 1400) who wrote that women do 'not have any set proportion' (Boyd, 1981, p. 202). Children, it seems, were too inconsequential for Cennini even to mention! As a reflection of progress in the art and scholarship of the time, the work of Leonardo and Dürer portended a major change in the concept of children and child-hood. In the year 1502, the physician Gabrielo de Zerbis described the anatomical differences between child and adult, and after 1600 the post-Renaissance painters began to depict children with normal proportions and also with growth pathologies. The Flemish artist Van Dyck depicts three normal children in the painting *The Children of Charles I* (1635). The painting *The Maids of Honor* by Diego Velázquez (1656) depicts a normal child, a woman with achondroplastic dwarfism (normal sized head and trunk with short arms and legs), and a man with growth hormone deficiency dwarfism (proportionate reduction in size of all body parts). Another post-Renaissance advance was a growing interest in how early life events could influence later development. For instance, by the 1700s physicians pursued the study of birth weight and its relation to child health.

The Count Philibert du Montbeillard of France measured the stature of his son every six months from the boy's birth in 1759 to his 18th birthday. Buffon included the measurements, and his commentary on them, in a

Supplement to his *Histoire Naturelle* in 1777. These data are usually considered to constitute the first longitudinal study of human growth. In longitudinal research, the same subject, or group of subjects, is measured repeatedly from year to year. A birth to maturity study may take upwards of 20 years to complete using this methodology, but the data may be analyzed for individual rates of growth and the timing of specific developmental events. Another type of research method, called cross-sectional research, measures subjects only once. Subjects of different ages may be included in a cross-sectional study to provide a general description of age-related growth changes, but the details of individual development, provided by the longitudinal approach, are lost. Buffon had earlier written on the adolescent spurt in gowth (the rapid acceleration in growth velocity around the time of sexual maturation) and on the general advancement of maturation of girls compared with boys. With the data on Montbeillard's son, Buffon noted the seasonal variation in rate of growth; the boy grew faster in the summer than in the winter. Buffon also wrote of the daily variation in stature; the boy was taller in the morning after lying at rest during the night than he was in the evening after working and playing during the day. Since Buffon's time, it has become necessary to take these variations in seasonal growth and daily stature into account when designing or analyzing longitudinal growth studies. Another 18th century longitudinal study of growth is that of the students of the Carlschule, conducted between the years 1772 and 1794. The pupils of this high school, founded by the Duke of Württenburg, included sons of the nobility and of the bourgeoisie. The growth data showed that the former were, on average, taller than the latter during the growing years but both groups achieved approximately equal height at 21 years of age. Thus, the sons of the nobility experienced an advancement of the rate of growth. This study, and the work of Buffon, clarify the important difference between amount of growth achieved at a given time and the rate of growth over time.

In 1835 Lambert Adolphe Quetelet published the first statistically complete study of the growth in height and weight of children. Quetelet was the first researcher to make use of the concept of the 'normal curve' (commonly called today the normal distribution or 'bell-shaped' curve) to describe the distribution of his growth measurements, and he also emphasized the importance of measuring samples of children, rather than individuals, to assess normal variation in growth. Quetelet's statistical approach was followed in Europe by Luigi Pagliani and in the United States by H. P. Bowditch. Pagliani began his studies on the size and fitness of Italian military personnel. He later applied his methods to children, and in 1876 demonstrated that the growth status and vital capacity (the maximum volume of air that can be inspired in one breath) of orphaned and abandoned boys, ages 10 to 19 years, improved after they were given care at a state-run agricultural colony. Pagliani also noted that children from the higher social classes were taller, heavier, and had larger vital capacities than poverty-stricken children. Finally, Pagliani followed Buffon in taking longitudinal measurements of the same children. From these he noted that menarche (the first menstruation of girls) almost always followed the peak of the rapid increase in growth that takes place during puberty. He concluded that reproduction was delayed in young women until growth in size was nearly finished. This, he considered, was a proper relationship, for the nutritional and physiological demands of growth would interfere with similar demands imposed by pregnancy.

Starting in 1875, Bowditch gathered measurements of height and weight, taken by school teachers, of 24 500 school children from the Boston, Massachusetts area. In a series of reports published in 1877, 1879, and 1891, Bowditch applied modern statistical methods to describe differences in growth associated with sex, nationality, and socioeconomic level between different samples of children living in New England. To account for the fact that children of the laboring classes were smaller than children from the non-laboring classes, Bowditch preferred an environmental, rather than a genetic, explanation. He said the non-laboring classes were taller because of the 'greater average comfort in which [they] live and grow up' (Boyd, 1981, p. 469).

This statement reflects the fact that during the 19th and early 20th centuries growth research was used for the first time to characterize the state of health of groups of children. Edwin Chadwick published findings in his *Report on the Employment of Children in Factories* that led to the passage of the Factories Regulation Act (1833) in England. The Act prohibited the employment of children under the age of nine and stipulated that periods for eating and rest must be provided for older children during the work day. The conditions of life for the factory children of Chadwick's time were horrible. Worse yet is that for more than half of the children alive today the conditions of life are little better. Undernutrition, poor health, illiteracy, and poverty are still rampant. Documentation of the pernicious effects of these conditions on the physical and mental growth of children continues to be carried out, and, in the tradition of Chadwick, recommendations for action to alleviate this suffering are made by researchers who have 'a feeling of responsibility for the children's welfare' (Borms, 1984).

In 1895 Franz Boas, an anthropologist working in the United States, published the first of a series of papers describing variability in growth and rate of development between populations of people, especially migrants to the United States and their children. Francis Galton had demonstrated, in his book *Natural Inheritance* (published in 1889), the hereditability of stature and other physical traits. Galton's work led some to believe that heredity was the all powerful determinant of human form and functional capabilities. One of Boas' contributions was the demonstration that the influence of the environment is as strong, or stronger, than heredity in controlling the expression of human physical characteristics. Boas found significant differences in physical features between adult migrants to the United States and their children born in the 'new country.' The children of migrants were almost always heavier and taller than their parents when both the children and the parents were measured at the same age. Boas ascribed this to the better health care and nutrition received by the children in the United States. Bowditch had found the same effect of migration on growth. Hereditarians believed that the ethnic origin of American-born children could easily be determined on the basis of physical measurements and that admixture between 'Anglo-Saxons' and people from southern and eastern Europe would bring about a physical degeneration of Americans. Bowditch proved statistically that the first of these contentions was not true, and he concluded that his research disputed the 'theory of the gradual physical degeneration of the Anglo-Saxon race in America' (Boyd, 1981, p. 469).

Though Bowditch started the anti-hereditarian campaign, it was the careful scientific approach used by Boas (he was trained in physics in Germany) that forced acceptance of the environmentalist position by all but the most staunch hereditarians. In 1911 Boas presented to the United States Congress a report titled 'Changes in the bodily form of descendants of immigrants,' which may have helped delay the imposition of limitations on migration from eastern and southern Europe. Despite the work of Bowditch, Boas, and other environmentalists, the passage of the Immigration Restriction Acts in 1921 and 1924 were victories for the hereditarians and eugenicists, since the acts specifically targeted southern and eastern Europeans for migration quotas (Gould, 1981).

Boas' scientific contributions include his research into the methodology of growth studies, which demonstrated the importance of calculating growth velocities from the measurements of individuals rather than from sample means. As shown in Figure 1.2, the former method gives an accurate estimate of average growth rate, while the latter method mixes data from early, average, and late maturing children and results in a mean velocity curve that underestimates the actual velocity of growth of all children during the adolescent growth spurt.

With the studies of Boas, the modern era of growth measurement and analysis began. In the first half of the 20th century several large-scale

Figure 1.2. (A) Individual velocity curves of growth (solid lines) and the mean velocity curve during the adolescent growth spurt. The mean velocity curve does not represent the true velocity of growth of any individual. (B) The same curves plotted against time before or after the age at maximum growth velocity of each individual. The mean curve accurately represents the average growth velocity of the group.

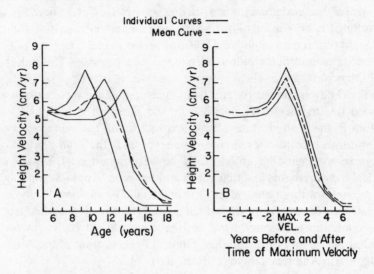

longitudinal studies of growth were started in the United States and Europe. Containing a wealth of data and information gained from advances in fields such as molecular biology, endocrinology, nutrition, and the social sciences, these studies allowed scientists and physicians alike to turn the study of human growth into a research and medical specialty. However, this was, primarily, descriptive research; that is, it told us how children grew and how their growth was affected by heredity and the environment, but it could not tell us why. To understand the why of growth and development a theoretical approach was needed. For example, the cell theory of Matthias Jacob Schleiden and Theodor Schwann, proposed in 1839, made it possible to understand the earlier work of Baer, who had described the different germ layers of the developing embryo. Without the cell theory it was impossible to know that the germ layers were distinct types of cells that gave rise to the different tissues and organs of the body.

With the publication of Darwin's *Origin of Species* in 1859 biological research became a modern theoretical science, and the scientific method of experimentation and hypothesis testing was increasingly applied to biological questions, including the control of growth and development.

One example is the work of Ernst Haeckle, who in 1891 proposed the theory of recapitulation during embryological development. This theory, that the development of the individual organism follows the evolutionary history of life, is now known to be incorrect in many details; e.g., mammalian embryos do not pass through a fish stage of development. Karl von Baer had earlier observed the similarities in appearance of embryos from different classes of animals (fish, reptiles, mammals, etc.) during early stages of development. Unlike Haeckle, Baer emphasized that during later development embryos from different classes move away from common forms, thus no recapitulation. Despite these differences, the work of Baer and Haeckle started the field of comparative embryology and was part of an intellectual trend that forced scientists to quantify their observations and propose testable hypotheses to explain their findings.

Many other scientists contributed to the growth theory during the late 19th and early 20th centuries, but the work of one person stands out more than any other. D'Arcy Thompson's book, *On Growth and Form* (1917; 1942), is a tour de force combining the classical approaches of natural philosophy and geometry with modern biology and mathematics to understand the growth, form, and evolution of plants and animals. Thompson visualized growth as a movement through time. Scientists from Buffon to Boas had studied the velocity of growth; Thompson made it clear that growth velocities in stature or weight were only special cases of a more general biological process. Thompson developed the concept and methodology of using transformational grids to quantify the process of growth during the lifetime of an individual or during the evolutionary history of a species (see Figure I.1). Until the advent of high speed computers, which are needed to carry out the mathematical procedure of the method, the transformational method was difficult and slow to apply and, hence, little used by other biologists. Even so, *On Growth and Form* provided an intellectual validity to growth and development research and stimulated succeeding generations of growth researchers to think about growth in new ways (e.g., see Huxley, 1932; Bookstein, 1978; Goldstein, 1984).

This brief review highlights some advances in the study of human growth. By the time of Boas and D'Arcy Thompson, many of the basic principles of biological growth and development were known. It was also acknowledged that all normal, healthy, and well-nourished children followed the same basic pattern of growth from birth to maturity. Research carried out during the last half century has shown that a common pattern of human development and growth occurs during the prenatal period as well. The remainder of this chapter describes the major features of human growth and development from conception to death.

Stages in the life cycle

Many of the basic principles of human growth and development are best presented in terms of the events that take place during the life cycle. One of the many possible ordering of events is given in Table 1.1, in which growth periods are divided into developmentally functional stages. For convenience, the life cycle may be said to begin with fertilization and then proceed through prenatal growth and development, birth, postnatal growth and development, maturity, senescence, and death. In truth, however, the course of life is cyclical – birth, the onset of sexual maturation in the adolescent boy or girl, and even death are each fundamental attributes of the cycle of life. In the child and adult, old cells die and degrade so that their molecular constituents may be recycled into new cells formed by mitosis. At the population level, people grow, mature, age, and die even as new individuals are conceived and born. Declaring one moment, such as fertilization, to be a beginning to life is arbitrary in a continuous cycle that passes through fixed stages in the individual person and in generation after generation.

Table 1.1. *Stages in the life cycle*

Stage	Duration
Prenatal life	
Fertilization	
First trimester	Fertilization to 12th week
Second trimester	Fourth through sixth lunar month
Third trimester	Seventh lunar month to birth
Birth	
Postnatal life	
Neonatal period	Newborn; birth to 28 days
Infancy	Second month to end of lactation; usually by age 24 months
Childhood	
Early	Milk-tooth period; second to sixth year
Middle	Permanent-tooth period; ages seven to 10 (approximate)
Later	Prepubertal period; from about age 10 to puberty, which typically occurs between 12 and 15 years in girls and 13 and 16 years in boys
Adolescence	The six years following puberty
Adulthood	
Prime and transition	Between 20 years of age and end of child-bearing years
Old age and senescence	From end of child-bearing years to death
Death	

Adapted from Timiras (1972).

food must be taken in to provide this energy, or the organism begins to break up.' To be sure, different tissues turnover at different rates, so that in muscle cells nitrogen is replaced in a few days to a few weeks, while the calcium in bone cells is replaced over a period of months. During the years and decades of life, sufficient turnover and renewal of the molecular constituents of the body's cells must take place to rejuvenate, virtually, the entire human being.

The metabolic dynamic of the human organism is most active during the first trimester of prenatal life. The multiplication of millions of cells from the fertilized ovum, and the differentiation of these cells into hundreds of different body parts, makes this earliest period of life highly susceptible to growth pathology caused by either the inheritance of genetic mutations or exposure to harmful environmental agents that disrupt the normal course of development (e.g., certain drugs, malnutrition, disease, psychological trauma, etc. that the mother may experience). Owing to these causes, it is estimated that about 10 per cent of human fertilizations fail to implant in the wall of the uterus, and of those that do so about 50 per cent are spontaneously aborted (Bierman *et al.,* 1965; Werner *et al.,* 1971). It is consoling to know, perhaps, that most of these spontaneous abortions occur so early in pregnancy that the mother and father are not aware that a conception took place.

By the start of the second trimester of pregnancy the differentiation of cells into tissues and organs is complete and the embryo is now a fetus. During the next 12 or so weeks most of the growth that takes place in the fetus is growth in length. During the first trimester, the embryo grows slowly in length reaching about 1.0 to 1.5 mm at 18 days after conception and about 30 mm, measured in terms of crown–rump length, at eight weeks after conception (Meire, 1986). By the fourth month crown–rump length is about 205 mm, by the fifth month 254 mm, and by the sixth month between 356 and 381 mm, which is about 70 per cent of average birth length (Timiras, 1972). Increases in weight during this same period are much less rapid. At eight weeks the embryo weighs 2.0 to 2.7 grams (O'Rahilly & Muller, 1986) and at six months the fetus weighs only 700 g, which is about 20 per cent of average birth weight (Timiras, 1972). It is during the third trimester of pregnancy that growth in weight takes place at a relatively faster rate. During the last trimester the development and maturation of several physiological systems, such as the circulatory, respiratory, and digestive systems, also occurs, preparing the fetus for the transition to extra-uterine life following birth.

Birth

Birth is the critical transition between life in utero and life independent of the support systems provided by the uterine environment.

The neonate moves from a fluid to a gaseous environment, from a nearly constant external temperature to one with potentially great volatility. The newborn is also removed from a source of supply of oxygen and nutrients provided by the mother's blood and passed through the placenta, which also handles the elimination of fetal waste products, to a reliance on his or her own systems for digestion, respiration, and elimination. The difficulty of the birth transition may be seen in relation to the percentage of deaths by age during the neonatal period (the first 28 days after birth). For example, Shapiro & Unger (1965) found, based on a study of all reported births in the United States during the first three months of 1950, that 49.5 per cent of all neonatal deaths occurred during the first 24 hours following birth. During the second 24 hours after birth the mortality rate dropped to 15.2 per cent, and it continued to drop each day so that by days 21 to 27 only 2.7 per cent of neonatal deaths occurred. Of course, most of these deaths were not due to the birth process itself, rather the leading factor associated with neonatal death was, and still is, low birth weight. An index of relative mortality by birth weight during the neonatal period is given in Table 1.2. Relative mortality is defined as the percentage of deaths in excess of the number that occur for infants within the normal birth weight range of 3.0 to 4.5 kilograms. These data are for infants at all gestational ages. Prematurity, defined as birth prior to 37 weeks gestation, may cause additional complications that increase the chances of neonatal death. However, an infant small for gestational age (i.e., of low birth weight), no matter what that age may be, is usually at greater risk of death than a premature child of the expected weight for gestational age (Gould, 1986).

Low birth weight, without prematurity, is the result of growth retardation during fetal life. The cause of this growth retardation may be congenital problems with the fetus, placental insufficiency, maternal undernutrition, or disease. However, most low birth weight, with or without prematurity, is associated with socioeconomic conditions. The incidence of low birth weight in the developed nations is 5.9 per cent of all live births; in the poorer developing nations the incidence is 23.6 per cent (Villar and Belizan, 1982). Even in the United States the socioeconomic

Table 1.2. *Index of relative mortality by birth weight (kg) during the neonatal period (Shapiro & Unger, 1965)*

Birth weight	1.0–1.49	1.5–1.99	2.0–2.49	2.5–2.99	3.0–4.49	4.5+
Relative mortality	86.1	33.0	7.9	2.0	1.0	2.2

relationship with birth weight is strong. When educational attainment is used to estimate general socioeconomic status, researchers found that 10.1 per cent of births to women with less than 12 years of schooling are low birth weight compared with 6.8 per cent of births to women with 12 years, and 5.5 per cent to those with 13 or more years of formal eduation (Taffel, 1980). Blacks and other non-white minority groups show consistent lower average birth weights compared with whites, and part of this difference is known to be due to the lower socioeconomic status of the non-white groups. However, when white and black women are matched for socioeconomic status, black women still give birth to a higher percentage of low birth weight infants. This suggests that ethnic or genetic factors also determine birth weight (Taffel, 1980). Though there is an hereditary contribution to birth weight, the sensitivity of fetal growth to the environment is manifest in the differences between rich and poor countries in the percentage of low weight births, the decrease in birth weight caused by smoking (Schell & Hodges, 1985; Garn, 1985) and alcohol consumption (Able, 1982), and the increase in low birth weight during times of famine (Stein *et al.*, 1975). Robson (1978) estimates the variance in birth weight due to fetal genotype to be 10 per cent, the variance due to maternal genotype to be 24 per cent, and the variance due to non-genetic maternal and environmental factors to be 66 per cent of the total variance. Because of the predominance of non-genetic factors, public health workers often use birth weight statistics as one indication of the well-being of a population.

Weight at birth is just one measurement that is commonly taken to indicate the amount of growth that took place during prenatal life. Recumbent length, the circumference of the head, arm, and chest, and skinfolds are others. Recumbent length is similar to stature, however the person measured is lying down and is stretched out fully by having the examiner apply pressure to the abdomen and knees. This can be measured at an age when stature (standing height) is still impossible to determine. The maximum distance between the vertex of the head and the soles of the feet constitutes the measurement. Circumferences measure the contribution made by a variety of tissues to the size of different body parts. For example, head circumference measures the maximum girth of the skull and hence, indirectly, the size of the brain. This is because of the intimate conformity between the brain and the tissues which surround and protect it, and the dominant role of the brain in determining head circumference. Similarly, arm circumference includes the measurement of bone, muscle, subcutaneous fat, and skin, but, for infants of the same size, variations in arm circumference are chiefly due to variations in the amounts of muscle and subcutaneous fat.

Some representative data on size at birth and at 18 years of age for several measures are given in Table 1.3. These average figures show that at birth boys are a bit longer, heavier, and larger headed than girls, but the girls have slightly more subcutaneous fat at birth than the boys. In reality there is such a wide range of variation in actual birth dimensions that the small average sexual dimorphism in size is biologically insignificant. At 18 years of age, in contrast, many men and women display well-marked sexual dimorphism in all of these growth variables except head circumference. Another difference between the infant and the adult is in body proportions. For children born in the United States, head circumference at birth averages about 70 per cent of length at birth. By age seven years head circumference averages 42 per cent of length and at maturity the average value falls to about 30 per cent. The reason for this change in percentage over time is that the growth of the brain proceeds at a faster rate than the growth of the body (Scammon, 1930). For the average child in the United States, head circumference reaches 80 per cent of mature size by about seven years of age, though length growth is only 68 per cent complete at the same age (Nellhaus, 1968; Hamill *et al.*, 1977). There are also proportional changes in the length of the limbs, which become longer relative to total body length during growth. The proportional changes are illustrated diagramatically in Figure 1.3.

The composition of the newborn's body in terms of adipose tissue, muscle tissue, and a variety of chemicals has been determined. The newborn's total body weight is about 12 per cent body fat and 20 per cent muscle mass. By adulthood, men average 15 to 17 per cent body fat and 40 per cent muscle mass; women average 24 to 26 per cent body fat and 35

Table 1.3. *Some measurements of size at birth and at age 18 for children born in the United States*

	Boys		Girls	
	Birth	18	Birth	18
Recumbent length (cm)[a]	49.9	181.1	49.3	166.7
Weight (kg)[a]	3.4	69.9	3.3	55.6
Head circumference (cm)[b]	34.8	55.9	34.1	54.9
Triceps skinfold (mm)	3.8[c]	8.5[d]	4.1[c]	17.5[d]
Subscapular skinfold (mm)	3.5[c]	10.0[d]	3.8[c]	12.0[d]

[a] Hamill *et al.*, 1977.
[b] Nellhaus, 1968.
[c] Johnston & Beller, 1976.
[d] Johnson *et al.*, 1981.

Figure 1.3. Diagram illustrating the changes in body proportions of human beings that occur during prenatal and postnatal growth (from Stratz, 1909).

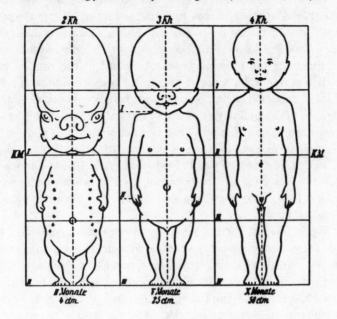

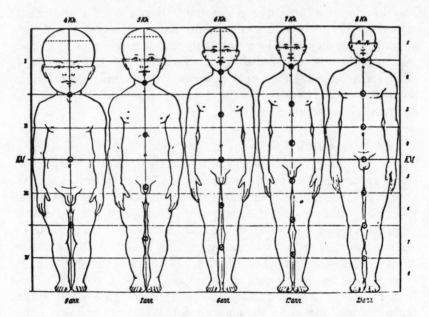

to 37 per cent muscle mass (Holliday, 1986; Katch *et al.*, 1980). The study of the formation and maturation of the skeleton during prenatal and postnatal life is another means towards describing different stages of development. Most bone forms from cartilage, which becomes calcified and, then, ossified into mature skeletal tissue. Bone formation takes place throughout the growing years. A record of the process can be captured on radiographs, since at certain X-ray exposure levels cartilage is 'invisible,' but calcified and ossified bone is radio-opaque (Figure 1.4). Radiographs of skeletal development of normally growing children from the United States and England have been compiled into atlases, which may be used to assess the stage of bone maturation of other children (Todd, 1937; Greulich & Pyle, 1959; Roche *et al.*, 1975a; Tanner *et al.*, 1983b). Data on chemical composition of the body (Widdowson & Dickerson, 1964) also show marked differences between values for newborns and adults.

The importance of these contrasts between early and later life is twofold. They allow clinicians and researchers to assess a child's stage of biological maturation for different organs, tissues, or chemicals independently of chronological age. Biological maturation is used to help determine if a child is developing too slowly or quickly, either of which may indicate the presence of some disorder. The contrasts between early and later life are also conceptually important. They show that the infant represents potentials for growth, maturation, and functional development. Because the human lifespan is long, relative to most other animals, these potentials are achieved after many years, during which a variety of factors may influence their final outcome. Adult human morphology, physiology, and behavior as well, are 'plastic' (Lasker, 1969) and in no way rigidly predetermined.

Postnatal life

The stages of postnatal growth and development, infancy, childhood, and adolescence span the time when newborn potentials achieve their mature realization. To visualize the amount and rate of growth that takes place during each of these stages, the growth in height of a child is depicted in Figure 1.5. The figure is based upon the most famous study of human growth, the biannual measurements of the son of the Count Montbeillard between the boy's birth in 1759 and his 18th birthday. The original data, reported in modern metric units by Scammon (1927), are drawn here as mathematically-smoothed curves (the cubic spline technique was applied to the data given by Scammon by the present author). The smoothing makes the important features of the curve more easily seen. The curve in part A of the figure is the boy's total height at

Figure 1.4. Radiographs of the hand and wrist illustrating the sequence of skeletal maturation events. (A) Newborn; no ossification centers present in the wrist, no calcified epiphyses. (B) Three years old; some carpal (wrist) ossification centers present, most epiphyses calcified. (C) Eight years old; all ossification centers calcified. (D) Thirteen years old; all bones have assumed final shape, but growth and epiphyseal closure remain to be completed (from Lowery, G. H.: *Growth and Development of Children*, 8th edn. © 1986 Yearbook Medical Publishers Inc., Chicago).

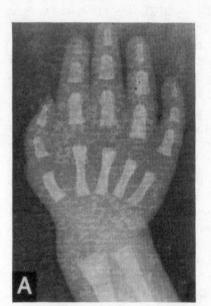
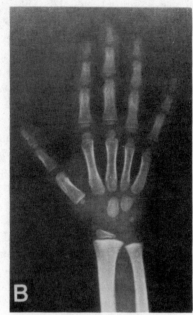
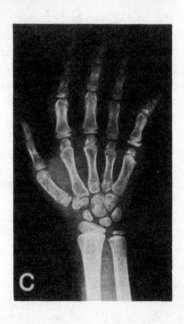
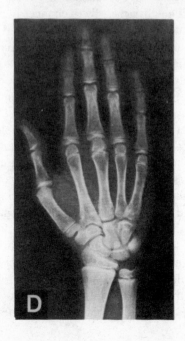

each measurement. If growth is viewed as a motion through time, then this graph may be called the distance curve of growth. The boy's rate of growth between successive measurements is graphed in part B of the figure, commonly called the velocity curve of growth. Each growth period is characterized by its own tempo of growth. The approximate divisions between periods are: infancy – birth to three years of age, childhood – three to 12 years of age, and adolescence – 12 to 18 years of age.

Growth during the infancy period is very rapid; during the first year of life infants may add 28 centimeters in length and seven kilograms in weight, which represents more than 50 per cent of birth length and 200 per cent of birth weight. The rate of decrease in velocity is also very steep, which makes growth during the first year more rapid than at any other time during postnatal life. The transition to childhood continues the deceleration in growth velocity, and by three years of age growth rate levels off to a slower, but steady, pace during the remainder of the childhood years. By the beginning of the childhood period, the youngster has achieved sufficient motor and sensory development to explore the environment and sufficient central nervous system (e.g., brain) growth to learn basic survival skills from this exploration. It has been argued that most of the childhood growth period is a time for learning about the physical, social, and cultural environment and that slow physical growth facilitates the efficiency of this learning (Huxley, 1932; Bartholomew & Birdsell, 1953; Dobzhansky, 1962).

The physical growth changes in height, weight, and body composition that take place during infancy and the preschool years are reviewed by Johnston (1986). He points out that one of the more striking features of human growth at this time is its predictability, both within individuals and between populations. The distance and velocity curves for height depicted in Figure 1.5 are examples of the predictability of childhood growth. Though this figure represents but a single child, the pattern of growth of all normal children follows a very similar course. For instance, Montbeillard's son, a boy of the French nobility raised in the countryside under near optimal conditions for that time, gained 59.9 cm in height between his second and 12th birthdays (Scammon, 1927). Children of, generally, middle socioeconomic class, born in the United States during the 1960s and early 1970s, average a 61.6 cm gain in height between their second and 12th birthdays (Hamill *et al.*, 1977). The difference between the gains in height of the French boy and the US sample is not statistically significant. The similarity in growth between a child and a sample of children, across time periods and across geographic boundaries, emphasizes the common pattern of growth shared by all normal children and the predictability of this pattern. These features of human growth

Figure 1.5. Distance and velocity curves of growth in height for Montbeillard's son (data from Scammon, 1927).

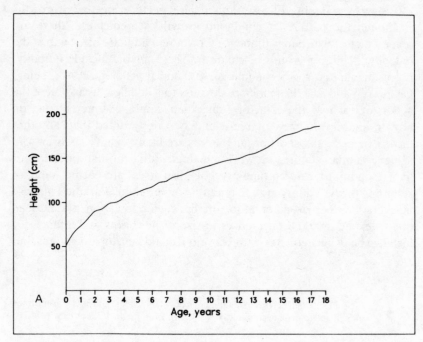

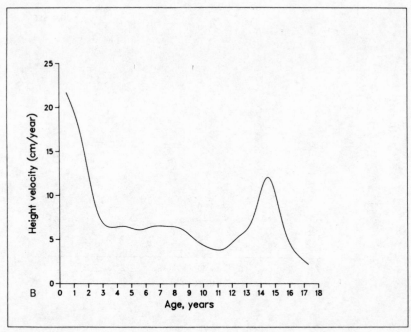

have important practical implications. For instance, they form the basis of epidemiological and clinical examinations that detect pediatric health disorders by searching for deviations in the expected trajectory of growth.

Though the pattern of childhood growth is predictable, there are several factors which may influence the amount and rate of growth of the individual child or groups of children. These factors include heredity, nutrition, illness, socioeconomic status, and psychological well-being. Chapters 4 and 5 of this book are devoted to a detailed discussion of the action of each of these factors, and their combined interactions, on growth and development. Here it may be briefly stated that, all other factors being equal, short or tall parents are likely to have children who achieve similar stature. However, malnutrition, chronic illness, poor living conditions, and chronic psychological stress are each capable of retarding the hereditary growth potential of any child. Variation in these hereditary and environmental factors between populations should lead one to expect marked differences in size of newborns and infants, for instance between neonates of the developed and developing nations, but

Figure 1.6. Mean heights of Guatemalan boys (solid lines) and girls (dashed lines) of high and low socioeconomic status (SES) (from Bogin & MacVean, 1978).

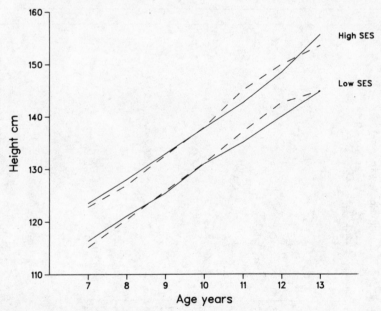

this is not the case. Habicht *et al.* (1974) and Van Loon *et al.* (1986) show that the growth of infants from a wide variety of ethnic and socioeconomic classes is remarkably similar during the first six months of life. Perhaps breast-feeding, which supplies the nutrient, immunity, and psychological needs of the infant, overrides the effect of variations in other aspects of the environment. By six months of age, when breast milk no longer meets the nutritional demands of the growing infant and other foods must be supplemented, children from the developed nations or higher socioeconomic classes are significantly larger than their less privileged age mates from poorer environments. With improved nutrition and health status early on, the disadvantaged children may catch up in size and reach their presumed genetic potential for growth (as shown by Pagliani in 1876). Otherwise, the differences in size between the well-off and the deprived become greater and greater, and by middle childhood the differences may become irreversible.

The growth in height and weight of Guatemalan children between the ages of seven and 13 years, shown in Figures 1.6 and 1.7, is an example of

Figure 1.7. Mean weights of Guatemalan boys (solid lines) and girls (dashed lines) of high and low socioeconomic status (SES) (from Bogin & MacVean, 1978).

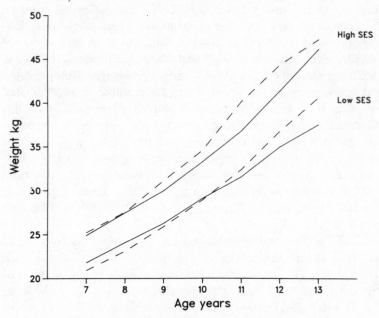

a stabilized difference in size. The larger children are from high socioeconomic status (SES) families, the smaller children are from low SES families. The high SES children are about the same size as healthy, well-nourished children from the United States (Johnston *et al.,* 1973). This fact, plus analyses of the nutritional status of the Guatemalan children, all point to the poor living conditions of the low SES children as the cause for their smaller size compared with high SES children (Bogin & MacVean, 1978; 1981a; 1983).

Though unequal in size, these children all display a similar regularity in their growth. As shown in Figure 1.8, children from both groups have, proportionately, the same height for weight regardless of absolute size. The regression line represents the 'best fitting' straight line (estimated by the statistical method of least squares) drawn through the data points for the high SES boys and girls. The data points for the low SES boys and girls show no statistically significant deviations from the regression line.

The maintenance of proportionality under the stress of low SES is another example of the predictability of childhood growth. Since both height and weight are equally affected in the low SES Guatemalan children, it is likely that some common mechanism is regulating the growth of several different tissues (e.g., bone, muscle, and adipose) and body parts (e.g., legs and trunk). The exact nature of this mechanism is not known. In the case of these low SES Guatemalan children, an insufficiency of calories and other nutrients and a high rate of infectious disease may be the factors that limit growth, but these factors do not explain why several tissues and body parts are proportionately reduced in size. This 'harmony of growth' (Widdowson, 1970) is also displayed in the normal development of many body parts, for example, the coordinated growth of the teeth and the craniofacial complex of bones that maintains the functional integrity of the masticatory system. Along with the phenomenon of 'catch-up' growth (Prader *et al.,* 1963) following short-term starvation or illness (i.e., a rapid increase in growth velocity which restores a child to his or her predicted size), the harmony and proportionality of childhood growth leads to a theoretical concept of growth as biologically self-regulating (Tanner, 1963; Goss, 1978; Bogin, 1980; and Chapter 7 of this book).

One other feature of the childhood phase of growth, noticeable in Figure 1.5, is the modest acceleration in growth velocity at about seven and eight years. This increase in velocity was noted in the analyses of growth published by Backman (1934), Meredith (1935), and Count (1943). Tanner (1947) called it the mid-growth spurt. Several recent studies have investigated the mid-growth spurt. In their London-based sample, Tanner & Cameron (1980) noted the presence of the mid-growth spurt in the average velocity curve of boys, but not girls (more than 150 children of each sex were measured). Molinari *et al.* (1980) found in a longitudinal study of children from Zurich, Switzerland (112 boys and 110 girls) that two-thirds of boys and girls had mid-growth spurts. Bock & Thissen (1980) found a mid-growth spurt in the longitudinal measurements of the 'average' child from a sample of 66 boys and 70 girls from Berkeley, California, and Meredith (1981) reported that 14 per cent of the 70 boys and 70 girls measured for the Iowa City growth study sample had the spurt. Berkey *et al.* (1983) found that 17 of 67 boys and none of the 67 girls from a Boston, Massachusetts growth study had mid-growth spurts. Varying methods of analysis, the number of serial measurements of height, and the frequency of measurements may explain the differences in findings between these studies.

The significance of the mid-growth spurt is not clearly understood. Some years ago, Bolk (1926) speculated that in our early human ancestors, sexual maturation took place at about six to eight years of age, but during our evolution the onset of sexual maturation became progressively delayed until reaching its present status as an adolescent event. In boys and girls the onset of sexual maturation is associated with a relatively large increase in growth velocity, the adolescent growth spurt. Current

Figure 1.8. Relationship of height to weight for Guatemalan boys and girls, aged 7.00 to 13.99 years old, of high and low socioeconomic status (SES) (from Bogin & MacVean, 1978).

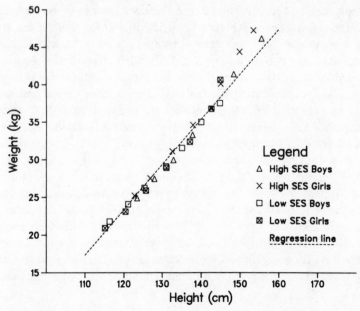

research links the mid-childhood spurt with an endocrine event called adrenarche (Albright *et al.,* 1942), the onset of secretion of androgen hormones from the adrenal gland. Some androgen hormones are believed to be able to promote faster growth in height. In boys, testosterone is an androgen secreted by the testes in relatively small amounts prior to puberty and relatively large amounts after puberty (Winter, 1978), and is probably the hormone that produces the adolescent growth spurt (Prader, 1984). Adrenal androgens may produce the mid-growth spurt in height. However, there is evidence that the major adrenal androgen, a hormone called dehydroepiandosterone sulfate, cannot produce increases in height growth velocity (Sizoneko & Paunier, 1982). Adrenarche generally precedes the onset of maturation of the gonads, called gonadarche, that takes place during puberty, and which does produce a large growth spurt. It is not known if adrenarche and gonadarche are directly related, though it has been proposed that both endocrine events have their origins in the maturation of brain structures, especially the hypothalamus (see review in Weirman & Crowley, 1986). Whether, in the distant past, gonadarche occurred at a young age simultaneously with adrenarche, as suggested by Bolk, is not known.

The rate of growth diminishes during childhood, reaching its nadir at the end of the late childhood period. The data for Montbeillard's son (Figure 1.5) show a dip in velocity at age 11. Similar dips are known from recent longitudinal studies of growth (Stutzle *et al.,* 1980), although not all children display them. The transition from late childhood to adolescence is marked by a reversal in the rate of growth, as may be seen by the shape of the velocity curve in Figure 1.5. The pattern of growth during the transition may also be expressed quantitatively as a change in the acceleration of height growth. Largo *et al.* (1978) analyzed the adolescent spurt in growth using data from a longitudinal study of children (112 boys and 110 girls) living in Zurich, Switzerland. The children were measured once a year, near their birthdays, between the ages of four and 18 years. This period of time is greater than the known variation in the timing of the onset of adolescence (occurring within the range of 6.6 to 13.5 years in this study), thus insuring that the transition from childhood to adolescence could be observed in every child. The relatively large number of children measured repeatedly from year to year gives confidence in the growth statistics derived from the study. Using the data published by Largo *et al.,* the change in acceleration of growth from childhood to adolescence can be calculated. During late childhood, height growth acceleration averages -0.46 cm/yr/yr for boys and -0.48 cm/yr/yr for girls, that is, growth rate is decelerating. From the point of minimal childhood velocity to the peak of the adolescent growth spurt the acceleration in height averages +1.66 cm/yr/yr for boys and +0.88 cm/yr/yr for girls.

The change in the velocity and acceleration of growth at adolescence affects almost all parts of the body, including the long bones, vertebrae, skull and facial bones, heart, lung and other viscera, and muscle mass. Some exceptions are that adipose tissue, both subcutaneous fat and the deep body fat, decreases in mass during adolescence in British and American boys, and perhaps in many girls as well (Tanner, 1965; Johnston *et al.*, 1974b). Lymphatic tissues and the thymus show no adolescent increase in size, indeed many of these tissues actually decrease in size during adolescence (Scammon, 1930). Changes in stature, muscle mass, and fatness that occur from childhood through adolescence are illustrated in Figure 1.9 for a longitudinally measured sample of French–Canadian children (Baughan *et al.*, 1980). Muscle mass is an estimate of the amount of muscle at the mid-point of the arm. This estimate is derived from measurements of mid-arm circumference and triceps skinfold. The circumference measures the contribution of skin, subcutaneous fat, muscle, and bone to the size of the arm. The triceps skinfold estimates the contribution of skin and subcutaneous fat to arm circumference. If it is assumed that the arm is cylindrical in shape, simple geometry may be used to calculate the lean arm circumference, which is the circumference of the muscle and bone at the mid-arm (Gurney & Jelliffe, 1973 give the formulae used to make this calculation). If it is also assumed that the circumference of the humerus is equal for all individuals, variation in lean arm circumference represents differences in the amount of muscle at this site. Though the arm is not cylindrical in shape and the circumference of the humerus is not equal in all individuals, the differences between reality and the assumptions of the technique are small enough so that when applied to groups of individuals, from developed and developing nations, the estimates of average muscle mass at the mid-arm are reliable and accurate (Jelliffe, 1966; Martorell *et al.*, 1976; Frisancho, 1981). Fatness is represented by the sum of three subcutaneous skinfolds: triceps, subscapular, and superiliac. It has been shown that subcutaneous and deep fat reserves are correlated and that changes in both occur in a similar fashion (Hunt & Heald, 1963), though recent studies using computed tomography to measure deep fat show a lack of correlation between the two fat reserves (Borkan *et al.*, 1982; Davies *et al.*, 1986). Nevertheless, since most fat is subcutaneous, a measurement of the amount of subcutaneous fat is a fair estimate of total fat (Brozek, 1960).

It may be seen in Figure 1.9 that there is little difference in average stature between boys and girls until adolescence, after which boys are

typically taller than girls. Girls usually begin their adolescent growth
spurt about two years earlier than boys, which means that in many cases
girls are taller than their male age-mates for a couple of years. Boys have
greater average muscle mass at all ages, though the differences become
absolutely greater, and biologically important, at adolescence. Con-
versely, girls tend to have more subcutaneous fat at all ages, and again,
the difference in fatness increases during adolescence. On average, girls
add fat mass continuously from age eight to 18; most boys experience an
absolute loss of total fat mass during adolescence, and may have no more
fat at age 18 than they had at age six (Holliday, 1986). The spurt in growth
of muscle mass in boys is usually accompanied by an increase in bone
density, an increase in cardio-pulmonary function, larger blood volume,
and greater density of red blood cells. Increases in each of these also occur
in girls, but at levels relatively and absolutely lower than for boys (Shock,
1966).

As indicated in Figure 1.5, the shape of the adolescent growth spurt is
not symmetrical. The rise to peak height velocity is relatively slower than

Figure 1.9. Mean stature, mean lean arm circumference, and median of the sum
of three skinfolds for Montreal boys and girls (from Baughan *et al.*, 1980).

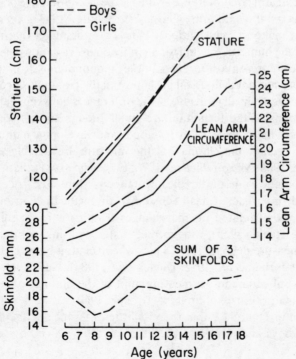

the fall after the peak. This has been observed for groups of children in recent studies (Largo *et al.*, 1978). The size of the spurt is greater in boys than in girls. Tanner *et al.* (1976b) found that boys have an average peak height velocity of 10.3 cm/yr; for girls it is 9.0 cm/yr. Largo *et al.* report similar results, though peak height velocity averaged only 9.0 cm/yr for boys and 7.1 cm/yr for girls. The size of the spurt and the age when peak velocity is reached are not related to final adult height, in fact children with certain endocrine disorders do not have a growth spurt but may reach normal adult height (Prader, 1984).

On average, adult men are taller and heavier than adult women. Alexander *et al.* (1979) surveyed 93 societies, including Western and non-Western cultures, and found that the stature of women averages between 88 and 95 per cent of the stature of men. In England, women average 93 per cent of the height of men, and this average difference is identical for men and women in the tallest (97 percentile), median (50th percentile), and shortest (3rd percentile) height groupings (Marshall, 1978). One study in Switzerland (Largo *et al.*, 1978) found that the difference between men and women in adult height is 12.6 cm. Since this study had followed the growth of the subjects longitudinally, from the age of four years, it was possible to calculate how much of the adult difference in height occurred in the various stages of postnatal growth. It was found that four factors accounted for the difference: the boys' greater amount of growth prior to adolescence added 1.6 cm, the boys' delay in the onset of adolescence added 6.4 cm, the boys' greater intensity of the spurt added 6.0 cm, and the girls' longer duration of growth following the spurt subtracted 1.4 cm from the final difference.

Owing to the interplay of these factors, the regulation of size may be more precisely controlled and the 'harmony of growth' evidenced during infancy and childhood is continued during adolescence. For instance, Boas (1930) discovered that the age at which adolescent growth begins is inversely correlated with the size of the spurt, meaning that early maturing children have higher peak height velocities than late maturing children. This observation has been confirmed for American children (Shuttleworth, 1937; 1939), British children (Marshall & Tanner, 1969; 1970), and Swedish children (Lindgren, 1978). Another compensating mechanism, described for Swiss children by Largo *et al.* (1978), is that a child with slow growth prior to puberty will tend to have a longer lasting growth spurt during adolescence than a child who achieves a greater prepubertal percentage of adult height. Where chronic undernutrition, disease, and child labor are endemic, such as in highland Peruvian Indian societies (Frisancho, 1977) and East African pastoral societies (Little *et al.*, 1983), height at every age is reduced compared with less stressed populations. However, the total span of the growth period is prolonged, up to age 25 or 26 years, so that a greater adult height may be achieved than if growth stopped at 18 to 21 years, as it does for healthy individuals in the United States. Presumably, these growth adjusting mechanisms are present in all children.

Sexual maturation takes place during the adolescent growth period. Ovarian maturation in girls and testicular maturation in boys result in the attainment of reproductive capabilities and many anatomical changes as well. Growth in height, weight, size of organs, and body composition have already been discussed. In both sexes, the onset of the growth spurt in height is followed within a few months by the appearance of the secondary sexual characteristics. In boys these include changes in size of the penis and scrotum, the growth of pubic, axillary, and facial hair, the 'breaking of the voice,' and seminal emission. In girls the secondary sexual characteristics include the growth of the breasts, appearance of pubic and axillary hair, and development of the uterus, vagina, and vulva to their mature size and appearance. Details of the development of the primary and secondary sexual characteristics, interrelationships between these events during adolescence, and sex differences in the timing of these events are available for American children (Simmons & Greulich, 1943; Reynolds & Wines, 1948; Nicolson & Hanley, 1953); for Chinese children (Lee *et al.,* 1963; Chang *et al.,* 1966); for British children (Marshall & Tanner, 1969; 1970; Billewicz *et al.,* 1981); for Swedish children (Taranger *et al.,* 1976); for Polish children (Bielicki, 1975; Bielicki *et al.,* 1984); and for Turkish children (Onat & Ertem, 1974; Neyzi *et al.,* 1975a; 1975b). All of these studies report findings for normal healthy children of middle to upper socioeconomic class.

Differences between samples exist in the timing of onset of the stages of adolescent maturation. For instance, breast development in American (of European descent) and British girls begins at an average age of about 11 years, which is about one year later than for Turkish or Chinese girls. Turkish boys begin pubic hair development at, on average, 11.8 years, compared with Swedish boys at 12.5 years and Chinese boys at 13.0 years. Even so, the results from each of these samples are remarkably similar, despite the variation in ethnicity, geographic areas, and cultural practices of each population, and variation in the methods of measurement and analysis used by the authors. The amount of variation in the age at which individual children achieve any maturational stage is greater within the samples studied than between the samples.

Nicolson & Hanley (1953) used factor analysis and Bielicki *et al.* (1984) used principal component analysis to describe their data. These statistical techniques divide the total variance in a set of data into discrete sources of variation, called 'factors' or 'components.' In both studies the factor accounting for the largest percentage of the total variance in maturation was a general maturity factor, on which clustered growth velocities for height and other linear dimensions, stages of sexual maturation (e.g., breast development in girls or pubic hair growth in boys), and skeletal maturation. For the data of Nicolson & Hanley, the general maturity factor accounted for an average of 71 per cent of the variance in maturation in boys, and an average 73 per cent of the variance in girls. Bielicki *et al.* found that the general maturity factor accounted for 77 per cent of the variance in boys and 68 per cent in girls. This statistical finding, along with the similarities in adolescent maturation found in different populations, supports the idea that adolescent maturation is controlled by some central organ or system within the body. These data also demonstrate that there is a human pattern of adolescent growth and development which is shared by all people.

Factors responsible for individual and population variation in adolescent growth and development are heredity, nutrition, illness, socioeconomic status, and psychological well-being, factors that were listed above in relation to variation in growth during childhood. The timing of menarche is, perhaps, the best studied adolescent event known to be affected by these factors (no similarly well-marked and dramatic event occurs for boys). From a study of monozygotic (MZ) and dizygotic (DZ) twin girls, Tisserand-Perier (1953) showed that the difference in age at menarche was 2.2 months for MZ twins and 8.2 months for DZ twins. MZ twins are genetically identical, while DZ twins share, on average, only half of their genes. Presumably, it is the genetic identity of the MZ twins that is responsible for their much greater concordance in menarchial age. The findings of this early study have been repeated recently by Fischbein (1977).

Nutrition, illness, and socioeconomic status are often linked together in human populations. For instance, poverty in the developing nations, and in the developed industrial nations, is almost always associated with high rates of protein–calorie undernutrition, high rates of infectious diseases, and illiteracy. Menarche is achieved at about 12.5 years of age in girls of the middle class from many nations, and at 14 years of age or later in girls of the lower socioeconomic classes (see reviews by Johnston, 1974 and Eveleth & Tanner, 1976). The latest mean age of menarche on record is 18 years of age for girls from the Bundi tribe of highland New Guinea (Malcolm, 1970). Malnutrition, heavy labor, and living at high altitude may contribute to their late menarche.

Finally, psychological factors influence growth and development, including menarche and menstruation (Ruble & Brooks-Gunn, 1982).

Malina *et al.* (1973) found that highly competitive female track athletes reach menarche later than girls in the general population. Frisch *et al.* (1980) found menarche is delayed in highly trained ballerinas. One explanation for the delay in menarche of these groups of girls is the stress of exercise. The production of several hormones, including progesterone, prolactin, and testosterone, is stimulated by strenuous physical activity. These same hormones are known to delay the onset of menstrual cycles (Scott & Johnston, 1982). Psychological stress may also be a cause of delayed menarche. Prima ballerinas are extremely sensitive about their weight and body image, and compulsive behavior in relation to these and their dance is often reported (Warren, 1980). Post-menarchial ballerinas may stop menstruating before a performance, even before they begin intense training or dieting (Scott & Johnston, 1982). Adequate studies to determine the psychological component of menarche and menstrual regularity have not been carried out; however, the existing data are suggestive.

Adulthood

The attainment of adult stature is one of the hallmarks used to mark the transition from adolescence to adulthood. In the United States, young women and men of middle to upper socioeconomic status reach adult height at about 18 years of age and 21 years of age respectively (Roche & Davila, 1972). Hulanicka & Kotlarz (1983) studied a sample of 221 young men from Wroclaw, Poland, an industrial city of 600 000 people, and found that only 54 per cent of the subjects reached final adult height by age 19 years. The other 46 per cent added an average of 2.13 cm in height between the ages of 19 and 27 years. It is well known that individuals suffering from undernutrition, chronic diseases, and certain drug therapies may continue to grow in height until they reach about 26 years of age. Though these individuals may grow for a longer period of time, they usually achieve less total growth and end up shorter than their more privileged or healthier age-mates.

Height growth stops when the long bones of the skeleton (e.g., the femur, tibia, etc.) lose their ability to increase in length. Usually this occurs when the epiphysis, the growing end of the bone, fuses with the diaphysis, the shaft of the bone. As in Figure 1.4, the process of epiphysial union can be observed from radiographs of the skeleton. In their study of Polish men, Hulanicka & Kotlarz (1983) found that the amount of growth that occurred after age 19 years was a function of skeletal maturation: late maturers grew more than average or early maturers. This fact has been known for many years, and an estimate of skeletal maturation, often called skeletal age, is incorporated into equations used to predict the adult height of children (Bayley & Pinneau, 1952; Roche *et al.*, 1975b; Tanner *et al.*, 1983a). The fusion of epiphysis and diaphysis is stimulated by the gonadal hormones,the androgens and estrogens. However, it is not the fusion of epiphysis and diaphysis that stops growth, for children without gonads, or whose gonads are not functional, never have epiphysial fusion, but they do stop growing (Tanner, 1978). Rather, it is a change in the sensitivity to growth stimuli of cartilage and bone tissue in the epiphysial region that causes these cells to lose their hyperplastic growth potential.

Reproductive maturity is another hallmark of adulthood. The production of viable spermatozoa in boys, and viable oocytes in girls, is achieved during adolescence, but these events mark only the early stages, not the completion, of reproductive maturation. For girls, menarche is usually followed by a period of one to two years of adolescent sterility. That is, there are menstrual cycles, which are often irregular in length, but there is no ovulation. So, the average girl is not fertile until 14 years of age or older. Regular ovulatory cycles, however, do not indicate reproductive maturation. Becoming pregnant is only a part of reproduction: maintaining the pregnancy to term and raising offspring to adulthood are equally important aspects of the total reproductive process. Adolescent girls who become pregnant have a high percentage of spontaneous abortions and complications of pregnancy. This is true for girls in developed nations such as the United States (Taffel, 1980) and developing nations such as Peru (Frisancho *et al.*, 1985). Teenage mothers also have higher rates of low birth weight infants than older mothers and, consequently, these infants suffer high rates of mortality (Taffel, 1980; Garn & Petzold, 1983). There are many reasons for the reproductive difficulties faced by teenage girls, ranging from physiological immaturity of the reproductive system to socioeconomic and psychological trauma induced by the pregnancy. The fact that the mother is still growing means that the nutritional needs and hormonal activity of her body may compete with and interfere with the growth and development of the fetus. This problem was suggested by Pagliani over 100 years ago and recently confirmed by Frisancho *et al.* (1985). For all these reasons, most researchers agree that female reproductived maturity is reached at the end of the adolescent stage of life.

Boys begin producing sperm at an average age of 14.5 years (Richardson & Short, 1978; Hirsch *et al.*, 1979; Laron *et al.*, 1980). Whether this event marks the onset of fertility is not known. The quality of viable sperm from teenage boys is also unknown, though one may speculate that pubertal endocrine activity in the boy may have some effect on his sperm cells. Even if fertile, the average boy of 14.5 years is only beginning his adolescent growth spurt and, therefore, his developmental status is incomplete. In terms of physical appearance, physiological status, and psychosocial development, he is still more of a child than an adult. For these reasons, the ability to successfully father children and care for them is usually achieved only late in adolescence.

Though the transition to adulthood may be marked by the cessation of height growth, reproductive maturity, and other physical and psychosocial events, the course of growth and development during adulthood is not easily described. There is a lack of precisely timed or sequenced physiological events. Most tissues of the body lose the ability to grow by hyperplasia, but many may grow by hypertrophy. Exercise training can increase the size of skeletal muscles and caloric oversufficiency will certainly increase the size of adipose tissue. However, the most striking feature of the adult stage of life is its stability (homeostasis) and its resistance to pathological influences, such as disease-promoting organisms and psychological stress (Dubos, 1965; Timiras, 1972). This contrasts with the preceding stages of life, characterized by change and a susceptibility to pathology.

Old age and senescence follow the prime years of adulthood. The aging period is one of gradual, or sometimes rapid, decline in the ability to adapt to environmental stress. The pattern of decline varies greatly between individuals. Though specific molecular, cellular, and organismic changes can be measured and described, not all of these occur in all people and rarely do they follow a well-established sequence. This suggests that, unlike the biological self-regulation of growth prior to adulthood, there is no biological or genetic plan for the aging process. There are many theories about the aging process and about why we must age at all. One theory with empirical experimental support links aging with the limited mitotic (cell duplication) ability of hyperplastic cells. Hayflick (1980) found that when raised in tissue cultures, human embryo hyperplastic cell lines double in number by mitotic division only 50 ($\pm$10) times and then die. Tissue cultures of cells from adult humans have an even more limited mitotic potential, doubling only 14 to 29 times before dying. This doubling limit of hyperplastic cells provides a theoretical limit to life. In practice, few people ever reach this limit. Rather, the inability of all cell types, including nerve, muscle, and other non-replicating cells, to use nutrients and repair damage begins before the cells die. Undoubtedly, aging is a multi-causal process. The reason there is no biological plan or developmental sequence for aging may be because there is no biological reason to age in any particular way. It is only recently, in the evolutionary history of our species, that human populations have come to live past the prime adult years. Throughout prehistory, death by predation, disease, and trauma caused by violence and accidents was probably more common than death due to old age (see Chapter 3). Death is inevitable, but nature did not have the time or the selection pressures to mold our manner of death into a predictable pattern.

In contrast to the process of aging and death, growth and development from conception to adulthood follow a predictable pattern. It was during the evolutionary history of our species, and those species ancestral to ours, that selective pressures operated to shape our pattern of growth. Thus, to understand why we grow the way we do, we must examine some of the events that occurred during human evolution. The next chapter describes the evolution of the human pattern of growth.

2 *The evolution of human growth*

Biologists and anthropologists have proposed a number of taxonomic schemes for classifying the uniqueness of *Homo sapiens*. Lovejoy (1981) recently suggested five characteristics of humans as defining features: bipedality, a large neocortex, reduced anterior dentition with molar dominance, material culture, and unique sexual and reproductive behavior. The development of each of these characteristics can be seen in the ontogenetic unfolding of the human pattern of growth and development. For instance, bipedality is made possible by differential growth of the legs and pelvis versus the arms and shoulder girdle. Our unique sexual behavior results, in part, from our prolonged childhood, delayed maturation, and species-specific neuroendocrine physiology. The human pattern of endocrine physiology results in the menstrual cycle of women, the continuous sexual receptivity of both sexes, and the development of our secondary sexual characteristics, e.g., patches of hair in the groin and armpits rather than fur all over. These are some of the unique features of *Homo sapiens* and human growth. Though these characteristics set us apart from all other species, they have their origins in evolutionary history. In this sense we share many basic growth patterns with other species, but differ through some special evolutionary developments. To understand better both the shared and unique features of human growth, the next sections of this chapter consider the phylogeny of growth of lower vertebrates, the mammals, and the primates – the group that includes monkeys, apes, and humans.

Vertebrate and mammalian foundations for human growth
The shape of the mathematical curve of growth of all organisms, parts of organisms, and colonies of cells is virtually identical. It is an S-shaped, or sigmoid, curve characterized by an initial acceleration and then a period of deceleration in growth rate (see Figure 2.1 A). The growth of bacteria, chickens, rats, cattle, and even tumors in animals follow this curve (Brody, 1945; Bertalanffy, 1960; Laird, 1967; Timiras, 1972). Figures 2.1 B and C illustrate some of the other mathematical features of the sigmoid curve. In B the velocity, or rate of growth, is given; only a single peak, or maximum rate of growth, occurs. In C the changes

42

in acceleration are revealed; the point of zero acceleration corresponds to the inflection point in the sigmoid curve where the rate of growth stops increasing and begins to slow. Growth rate at any subsequent point on the curve is decreasing with time.

Figure 2.1. General growth curves: (A) weight versus time, (B) rate of growth (velocity) versus time, and (C) acceleration of growth rate versus time (after Medawar, 1945).

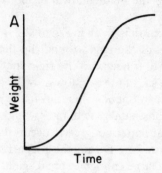

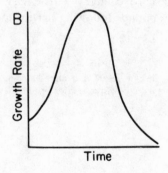

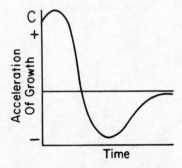

In mathematical terms, parts B and C of this figure are related to A as its first and second derivatives, respectively. The fact that the biological pattern of growth can be so clearly described mathematically is fortunate. That the curve of general growth is completely differentiable means, mathematically, that the growth process that this curve represents is a smooth and continuous one. Regular changes in amounts and rates of growth allow us to predict the course of development with precision and allow us to make quantitative and qualitative comparisons between different species of animals in terms of the mathematical properties of their growth curves.

An example of this type of growth pattern is given in Figure 2.2 for the chicken. Only the physical constraints of the egg, around the time of hatching, interfere with a smooth growth trajectory. The rigid shell and the depletion of nutrients from the yolk sac of the egg slow growth before hatching. After hatching the growth rate 'rebounds,' but only to the point where an averaging of the prenatal and postnatal growth rates would yield a smoothly decelerating curve. A similar pattern of growth occurs during the perinatal period of humans (see Figure 2.3). During the last part of the third trimester of pregnancy the fetus is large enough to press against the inner surface of the uterus and the placenta, which probably constricts blood vessels and inhibits the feto–maternal exchange of nutrients, gases, and wastes. Fetal growth slows, but rebounds following parturition so

Figure 2.2. Rate of growth of the chicken before and after hatching. The interrupted line is the theoretical curve if no growth restriction prior to hatching takes place (from Timiras, 1972).

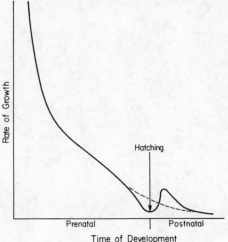

that the child 'catches-up' to the size he would have achieved if there had been no prenatal decrease in growth rate (Tanner, 1978).

Bertalanffy (1960) showed that the growth of mice and Brahman cattle may be modeled with the same curve as that used for the chick. Thus the pattern of growth of these phylogenetically and ecologically distinct organisms is qualitatively identical. The major exception to the general pattern of organic growth is the one followed by humans. The human pattern is illustrated in Figure 2.4, which is based upon data from a Swiss longitudinal study of growth (Prader, 1984). The curves are drawn from median values of growth, and thus represent growth of the typical boy or girl from the Swiss study. The curves in Figure 2.4 A, which represent

Figure 2.3. Distance (A) and velocity (B) curves for growth in body length in human prenatal and early postnatal life. Diagrammatic, based on several sources of data. The interrupted line is the theoretical curve if no uterine restriction takes place (from Tanner, 1978, *Fetus into Man*, Harvard University Press; reprinted with permission).

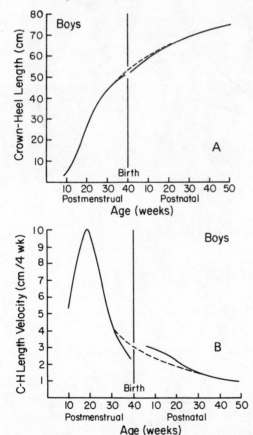

total height at each measurement (the so-called distance curve of growth) are, at first glance, not markedly different from the general sigmoid curve of Figure 2.1 A. However, the velocity curves illustrated in B are different from the velocity curves for other animals (Figures 2.1 B and 2.2). For the human there is a rapid decrease in growth rate from birth to about the age of four, followed by a more gentle decrease in the rate of growth to about age 10 for girls and age 12 for boys. Between the ages of approximately 10 and 12 for girls and 12 and 14 for boys there is a rapid increase in growth rate. This is the adolescent growth spurt, a unique feature of human growth in terms of its intensity and duration. Finally there is the constant decrease in growth rate that ends with the attainment of adult stature. Thus, qualitative and quantitative differences exist between human growth curves and their non-human counterparts.

Figure 2.4. Growth in height of typical boys and girls: (A) distance curve of growth achieved versus age, (B) velocity curve of growth rate versus age (from Prader, 1984).

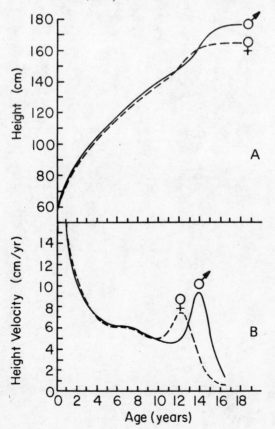

Unlike the non-human curve of growth, the human growth curve cannot be modeled with a single mathematical function, that is, it is not completely differentiable over its length. The distance curve requires at least two functions, one for the pre-adolescent segment and one for the adolescent segment (Shohoji & Sasaki, 1984). The velocity curve requires at least three mathematical functions for adequate description, one for the infantile segment (birth to about four years), a second for the childhood segment (from about four years to the beginning of the adolescent spurt), and a third for the adolescent segment (Laird, 1967; Bock & Thissen, 1976; Bogin, 1980; Karlberg, 1985). These mathematical features of human growth are unique, but they have their origins in the patterns of growth followed by mammals, in general, and the primates, in particular.

Mammalian growth

The growth of mammals differs from that of other vertebrates for two basic reasons. One reason relates to mammalian locomotion and the other relates to mammalian reproduction. These animals evolved the capabilities for rapid and flexible movement. This requires muscle tissue and something for it to work against. Vertebrates utilize bone for this purpose, a tissue that provides support and protection owing to its rigidity but also the developmental flexibility that allows for growth (Goss, 1978). Bone is found in fish, amphibians, and reptiles, but these animals make poor use of it compared to mammals. The bone of fish, amphibians, and most reptiles grows by periosteal deposition, the addition of tissue on all of the external surfaces of the bone. In a long bone, for instance, this means that as the bone grows it not only elongates at its articular surfaces but it also widens along the shaft. In these animals, bone growth never completely stops. Though growth continues very slowly after sexual maturity is reached, the prodigious size attained by dinosaurs and giant Galapagos tortoises may be explained, in part, by continuous growth over a long lifetime.

This pattern of growth is unsuitable for the mammals. The physiological characteristics of mammals, including homoiothermy (self-regulation of a relatively constant body temperature), efficient placentation by which the fetus continuously benefits from the maternal blood circulating in the uterus, rapid bodily movement, and other features associated with a relatively high metabolic rate, require a diet rich in energy and other nutrients. In the long term, mammalian metabolism requires a constant and high quality dietary intake rather than episodes of abundant food. Mammals must be able to move rapidly and efficiently to find and capture quality foods on a regular basis. This requires an efficient musculo-skeletal system during both the early and the later phases of growth. To maintain efficiency of function, a pattern of limited growth and remodeling of bone evolved to meet mammalian needs for movement and diet.

Some reptiles and most mammals have put an end to unlimited growth through the evolution of the cartilaginous growth plate system. A diagram of a typical mammalian long bone, its diaphysis, epiphysis, and growth plate, is given in Figure 2.5. The growth plate separates the growing part of the bone from the rigid part. This allows bone fully to perform its functions throughout growth and also allows for the cessation of growth. The latter is necessary for terrestrial animals that must support their own body weight without the help of water or any other buoyant medium. An end to growth is also necessary for terrestrial mammals that depend on rapid and flexible movement to find food and avoid predators. The largest terrestrial mammals, the Proboscidea (elephants and the allied extinct mammoths and mastodons), may have

Figure 2.5. (A) Diagram of a limb bone with its upper and lower epiphyses. (B) Diagrammatic enlargement of the growth plate region: new cells are formed in the proliferating zone and pass to the hypertrophic zone to add to the bone cells accumulating on top of the bone shaft (from Tanner, 1978, *Fetus into Man*, Harvard University Press; reprinted with permission).

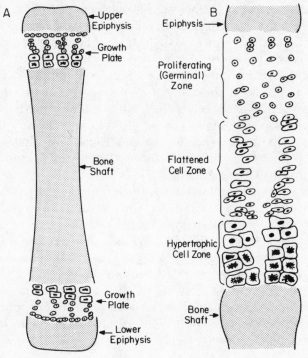

reached the limits of size for land animals of the class. The limbs of these animals are used almost entirely for support of the body and locomotion. The evolution of a flexible muscular appendage, the trunk, serves the function of a limb for food gathering and environmental manipulation.

Some mammals, such as rats, never stop growing; their growth plates never fuse with the diaphysis. However, they are capable of flexible and rapid movement throughout their lives. These animals grow so slowly and die so soon after sexual maturity that they never attain the sizes suggested by certain second rate horror films (though one rodent, *Hydrochoerus hydrochaeris,* the capybara of South America, may attain 1.3 meters in length and 50 kg in weight).

Another fundamental aspect of mammalian bone is its remodeling during growth. As a bone grows in length or size its surfaces must be reworked so that its characteristic shape and function can be retained. In a long bone this remodeling is achieved by removing old ossified bone tissue from the periosteal (outer) surface and adding new bone tissue to the edosteal surface (surrounding the hollow or sponge-like core) (Enlow, 1963; 1976). This process is schematically illustrated in Figure 2.6. Mammals achieved the efficient, rapid, and flexible mobility that they require via bone remodeling, the cessation of growth of the skeleton, and evolutionarily derived functional alterations in the articulation of limb bones (Romer, 1966).

Reproduction is the second aspect of mammalian biology that influences growth. Mammalian reproduction may be viewed as an adaptive strategy to provide a higher quality of parental care of offspring than that generally provided by non-mammals, which increases the reproductive success of the parents by decreasing pre-reproductive mortality of the offspring. Mammals were not the first to evolve parental care and investment in their offspring, but they have carried it to a level of physiology and behavior exceeding that of other vertebrates. The quality of mammalian parental investment may be measured by the efficient internal fertilization and placentation of most mammals, lactation by the mother during her offspring's infancy, and in the capacity of each individual offspring to help insure its own survival to reproductive age. Each of these mammalian features has a direct relationship to growth.

The evolution of the placenta removed some of the limitations to prenatal growth, including both growth in size and length of gestation. The prenatal growth and gestation of non-placental animals such as most reptiles, birds, monotremes (the platypus and echidna), and most marsupials (e.g., opossums and kangaroos) is limited by the need to 'package' fetal nutrients in the yolk sac and fetal waste products in a separate sac called the allantosis. In contrast, the placenta provides for fetal nutrition, respiration, and the removal of metabolic wastes continuously throughout gestation.

Mossman (1937) described the placenta as 'any intimate apposition or fusion of fetal organs to the maternal tissues for physiologic exchange.' Though the contact between tissues may be 'intimate' there is never any direct connection between mother and fetus and the exchange of substances always occurs through a tissue boundary. Amoroso (1961) described three basic kinds of placentae for mammals (Figure 2.7). The first was the yolk sac placenta, found in some marsupials and rabbits, in which blood vessels connect the yolk sac with the uterine wall. The second was the chorionic-allantoic placenta, characteristic of higher mammals, in which parts of the surface of the allantosis fuse with the chorion (a membrane surrounding the fetus composed of maternal and

Figure 2.6. Diagrammatic representation of remodeling in a limb bone. Level AA′ becomes repositioned into level BB′ as a result of the increased growth in length of the entire bone. The relative level of AA′ in the larger bone is indicated by an X. As a result of the relocation of AA′ to BB′, note that: (1) the sectional shape and (2) the sectional diameter have changed. Note also that the point indicated by the black arrow has been relocated from the inner side of the cortex in AA′ to the outer side of the cortex in BB′. All of these changes involve structural remodeling by the process of resorption (− signs) and deposition (+ signs) (from Enlow, 1963, *Principles of Bone Remodeling*; courtesy Charles C. Thomas, publisher, Springfield, Illinois).

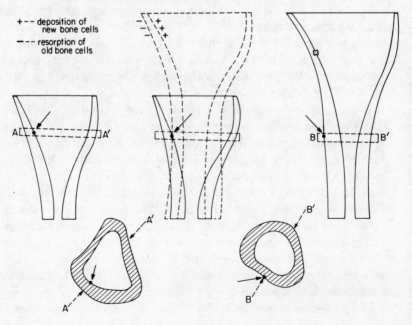

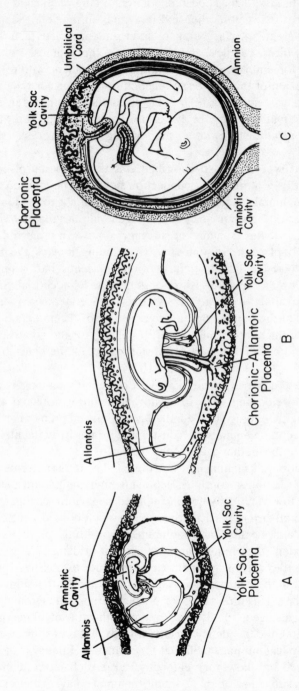

Figure 2.7. (A) The yolk sac placenta. (B) The chorionic–allantoic placenta. (C) The chorionic placenta. See text for details (after Hamilton & Mossman, 1972).

embryonic tissues). This apposition allows for a more efficient exchange of substances between mother and fetus than in the yolk sac placenta, because a greater surface area of fetal and maternal tissues are in contact. In birds, reptiles, and monotremes the allantosis serves as a receptacle for embryonic wastes during development within the egg. The conversion of this waste sac function to a system for the exchange of nutrients, gases, and wastes in some placental mammals is an example of the conservative nature of evolution. Existing organs are often 're-tooled' for new functions rather than having organisms develop totally new organs.

The third type Amoroso described was the chorionic placenta, in which there is a more direct connection between the chorion and the fetus via the umbilical cord. The chorionic placenta is found in rodents, monkeys, apes, and humans. There is generally a greater amount of surface area for the exchange of substances between fetus and mother in this type of placenta. In some species, including humans, the maternal and fetal blood circulations in the chorionic placenta share a single wall of tissue between capillaries (this is also called a hemochorial placenta). Such close contact over a relatively large surface area – gross external dimensions of the human placenta average 16 to 20 cm in diameter and 3 to 4 cm in thickness (Timiras, 1972) – allows for the greatest ease of diffusion and active transport of substances across the placenta of any mammal.

The critical advance in the biology of placental mammals is that the fetus can develop and grow to an advanced stage protected and well nourished in the uterus. The length of gestation and prenatal growth (in size) of placental mammals are constrained primarily by the mechanical limitations of the mother's uterus and birth canal.

By definition all mammals, even the egg-laying monotremes, nurse their young. Lactation continues supplying high quality nutrients to the newborn. However, its evolution required behavioral changes in both mother and offspring, particularly mother–infant bonding, which maintains the infant in contact and communication with the mother so that it can be suckled when hungry and protected if in danger. The mother–offspring contact ensuing from this feeding method establishes a period of dependency in the young and a reciprocal period of parental investment by the mother. This time of life for the newborn is called infancy, and it has become a stage in the life cycle and growth curve of all mammals. A similar period of dependency occurs for birds, but is of shorter duration and less physical intimacy than that for mammals. Infancy prolongs the growth period and allows for greater physical and behavioral adaptability. Compared with other animals, infant mammals may be better able to adjust total rates of growth, or the rates for specific body parts, to adapt to environmental stress. For example, in cold environments, growth rates may be adjusted to produce adult mammals with relatively larger bodies, shorter extremities, or both, compared with mammals living in warmer climates. Large bodies with short extremities conserve heat better than smaller bodies with relatively long extremities; the former body type has relatively less surface area for heat dissipation. The body size and body proportions of polar bears (*Thalarctos maritimus*) and Malayan sun bears (*Helarctos malayanus*) conform to these growth adaptations. Infancy may also increase behavioral adaptability by allowing young mammals the time to practice and improve innate behaviors, such as the stalking of prey by carnivores. The mother–infant bond increases the opportunity for young mammals to acquire learned behaviors by observing and imitating their mothers or other adult animals with whom the mother socially interacts.

The young mammal's capacity for learning, and the infant growth period that is its basis, relate to the last of the mammalian characteristics listed above, namely the ability of each offspring to help insure its own survival to reproductive age. The way mammals accomplish this is through the growth of relatively large brains and the flexibility in behavior that these large brains allow. The evolutionary record shows that the mammalian brain has undergone repeated selection for increases in size and complexity. Jerison (1973) compared endocasts (molds of the interior of the skull which may be used to estimate brain size) of fossil skulls of mammals and found that the brains they contained were, in proportion to body size, smaller in earlier times and have increased in size steadily over the last 60 million years. In contrast, Jerison found that reptiles have not undergone this selection, the brain size to body size ratio of reptiles has not changed appreciably during the last 200 million years.

Mammals have also evolved more complex and functionally diversified brain structures. The mammalian neocortex and its neurologically distinct regions (the motor–sensory region, the auditory and visual regions, etc.) are examples. Jerison (1976) showed that mammalian brains have a system of neurological pathways that bring together, at various locations, information from the visual, auditory, and olfactory senses. The 'integrative neocortical system,' as Jerison called it, joins neurological regions of the paleocortex (including the olfactory bulb and the limbic system) and the neocortex (including the visual, auditory, and somatic systems) of the brain (Figure 2.8). Lower vertebrates, such as reptiles, rely mostly on the paleocortex for the control of behaviors which Jerison characterized as 'fixed-action patterns . . . with few requirements for plasticity or flexibility' (1976, p. 101). Higher vertebrates, the birds and mammals, rely on both the paleocortex *and* the neocortex for the control of behaviors which are plastic and flexible in all species. Birds do not have the integrative neocortical system and, according to Jerison, their behavior displays to perfection the fixed-action pattern of response to stimuli (Jerison did not address the fact that birds can learn quite complex and lengthy behaviors). Mammals have the integrative system, allowing 'sensory information from various modalities [to be combined] as information about objects in time and space' (1976, p. 101). Mammals do not just react to environmental stimuli, they perceive, store, retrieve, and evaluate information and adjust behavioral responses according to the present situation and past experience. Mammals require more neurological tissue to accomplish these sensory, brain, and behavioral tasks and mammals do have brains that are larger, in proportion to body size, than the brains of reptiles and most birds. Larger, more complex brains allow for a greater capacity for learning and more flexible behavior, because learned behavior may be constantly modified by further learning.

Figure 2.8. Schematic view of the neurological connections between the visual, auditory, and olfactory systems of living mammals. The arrows show the general direction of the flow of information through successive orders of nerve cells (I through V; IV$_R$ indicates parts of the reflex control systems; F indicates feedback loops). The integrative system (right) is a mammalian characteristic (from Jerison, 1976).

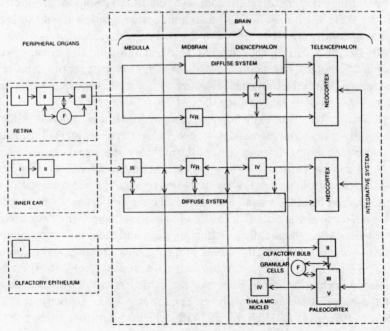

The evolution of learning as an adaptive strategy is associated, in a classic feedback manner, with the series of changes in mammalian biology and behavior that have just been described. The tissues of the central nervous system have relatively high metabolic activity, requiring a regular supply of nutrients and oxygen for maintenance and growth. The evolution of the placenta is directly related to the evolution of larger, more complex brains and greater learning abilities. The placenta is the organ that constantly supplies oxygen and nutrients to the developing fetal brain and allows that brain to develop to an advanced stage before birth. The evolution of the placenta and of the brain are clearly correlated in mammals. Yolk sac placentae are found in the lower mammals, chorionic–allantoic placentae are found in higher mammals, and chorionic placentae are found in the primates. In turn, each higher mammalian group has a brain that is relatively larger than expected for its body size. It is no coincidence that humans have one of the largest brain to body weight ratios, perhaps the most complex and active brain of any mammal, as well as the most efficient placental system (the hemochorial variety of chorionic placenta) of any primate.

Other biological changes related to brain evolution occur during postnatal life; one example is lactation, which was discussed above. In this feeding relationship there is a correlation of lactational behavior of the mother with brain size and learning. Martin (1968) described such a case for the tree shrew (*Tupaia belangeri,* a member of the order Insectivora), an animal with a moderate brain to body weight ratio and average learning ability compared with other mammals. Female tree shrews may cache their infants (two or three are born per litter) in a nest and leave them for up to 48 hours while they search for food. The infant tree shrews are virtually silent and unmoving during their mother's absence, which may be a behavioral adaptation to avoid attracting predators. The seclusion and immobility also limit the variety of sensory stimuli that the infants experience. Upon the mother's return the infants are nursed with a milk that is concentrated in calories and other nutrients. This pattern of periodic feeding coupled with sensory deprivation during infancy works well for a species with limited brain growth after birth and a limited learning potential. This feeding style would not work for a species with rapid postnatal brain growth, requiring a constant nutrient supply during infancy, and a greater dependence on learning in later life.

In neurologically more advanced mammals, especially primates, mother and infant remain in virtually constant physical contact for several weeks or months after birth. The norm for primates is one infant per pregnancy, which facilitates intimate physical contact since there is no competition between siblings for the mother. Suckling is done 'on demand,' 24 hours per day. The concentration of nutrients in the milk of primates is lower than that of the 'primitive' mammals, but the efficiency, constancy, and quality of nutrient supply is superior (Widdowson, 1976). The newborn primate is highly active compared with the tree shrew infant. The primate infant travels with its mother, sensing many of the things the mother experiences and developing motor and sensory skills in the process. This type of early sensory stimulation is known to be conducive toward further learning (Jolly, 1985). The infant primate grows more slowly than most other mammalian newborns and, therefore, is dependent for a longer time on this intimate relationship with its mother. Infant dependency extends the period of growth, development, and protection and also increases the opportunity for the infant to learn survival skills by observing successful maternal behaviors.

Infancy, dependency, and learning are advantageous to both mother and infant since they lead to a greater probability of survival to reproductive age of the young. The drawback of infant dependency is that it is incompatible with adult behaviors, particularly reproductive behavior. Competition for nesting space and breeding territories, aggressive encounters with conspecifics for mates, and mating itself are precluded behaviors for dependent young. Thus, sexual maturation of the offspring must be delayed until the dependency period is terminated. The mother's reproductive behavior is also curtailed during the dependency period. The delay in sexual maturation of the young and limits to total fertility of females are partially offset by the higher quality of mammalian reproduction. That is, some non-mammalian species, such as many kinds of insects, fish, and reptiles, rely on prodigious egg production to assure the survival of some of their offspring to reproductive age. In contrast, mammals maximize the probability of survival of each individual offspring to achieve a high degree of reproductive success.

Although all mammals share these basic reproductive adaptations and growth patterns, some mammals have evolved a pattern of rapid growth during prenatal or early postnatal life that leads to a shorter dependency period and the relatively early onset of sexual maturation. This strategy, called precocial development, is found in rodents like the rat and hamster. Total fertility may be increased by this strategy, but at the expense of parental investment in any given offspring or litter of young. The opposite strategy, called altricial (relatively delayed) development, is found in most of the social carnivores, for instance, wolves and lions (Walker, 1975). It is important to note that for most precocial and altricial mammals, the timing of sexual development takes place at the same relative point on the growth curve. It is at this point, where growth is still taking place but the rate is decelerating rapidly, that infancy ends and adulthood begins with no noticeable transition in growth rate. The most notable exception to this general pattern of development is the course of growth followed by the primates.

Primate growth patterns

Primates have a juvenile stage of development which occurs between infancy and adulthood. Juvenile mammals may be defined as 'prepubertal individuals that are no longer dependent on their mothers (parents) for survival' (Perieira and Altmann, 1985, p. 236). Most non-primate mammals grow in size and develop toward sexual maturity along a smooth and continuous path from birth to adulthood, with no biologically or mathematically discernible alterations in growth trajectory. The old world monkeys, apes, and humans follow a pattern of growth that differs from most other mammals in two major ways. The first is that sexual development is deferred until a time well after the infancy period of postnatal growth takes place. This period of delayed growth is the primate juvenile growth period. The second difference is that neurological development, especially growth of the brain, is about 90 per cent complete before sexual maturity is achieved.

Laird (1967) reviewed the growth of the rhesus monkey, chimpanzee, and human, the species that have been most extensively studied. She found that monthly weight increases in the male rhesus monkey (Figure 2.9) followed two separate growth curves, the first from birth to 22 months and the second from 23 months onwards. The curves were fitted

Figure 2.9. Growth in body weight of the rhesus monkey (from Laird, 1967).

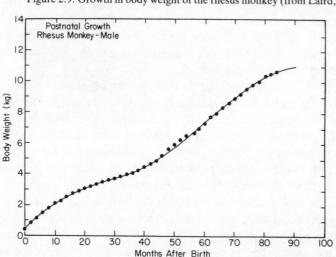

to the monthly weight data points by the method of least squares regression. Essentially, this method minimizes the sum of the squared deviations of each data point from the fitted curve, i.e., it produces the 'best-fitting' line between the data points. Using this method, Laird found that the change in growth rate that occurred between the two curves, between months 22 and 23, created the need to use different mathematical functions to model growth in the two periods. Sexual development takes place during the second growth phase, after month 40. The deviations of weight growth above the fitted curve between months 48 and 54 corresponded to the time of reproductive maturation and the beginning of adult levels of gonadal hormone secretions. The curves of growth for the female rhesus were similar, the time of sexual maturation was earlier though, occurring at about 42 months.

Laird found that the growth of the chimpanzee (Figure 2.10) also required two mathematical functions to describe its course from birth to maturity. The early phase of growth was best approximated by a linear function from birth to six years. This was followed by an adolescent phase that was modeled by a curvilinear function. Male and female chimpanzees followed the same pattern of growth from birth to adulthood. During the first phase the amount of growth achieved by males and females was identical, but during adolescence sexual dimorphism in size became well marked; male weight growth deviated from the fitted curve more than female growth. The deviations were greatest at the time of complete sexual maturation (the age at which male chimpanzees begin to sire offspring) and were closely associated with the onset of adult levels of gonadal hormone secretions (Martin *et al.*, 1977). The changes in weight growth increments are represented in Figure 2.11 as velocities. Only the male showed a noticeable 'spurt' in weight velocity. In other studies, Gavan (1971) and Watts & Gavan (1982) found that in contrast to weight, growth velocities of the long bones and the trunk of the body showed no visually detectable spurt in either sex.

Laird (1967) found that the velocity curve of human growth, as depicted in Figure 2.4, required three mathematical functions to model its course, a conclusion also reached, independently, by Bock & Thissen (1976), Bogin (1980), and Karlberg (1985). The need for the third function is one aspect of human growth that makes it unique, even among the primates. Laird described the similarities of growth between primate species and the uniqueness of human growth as follows:

> the curvilinear growth by which the body weight of an organism approaches its mature value and during which sexual maturation characteristically occurs, starts at birth in sub-primate mammals and birds, but is deferred in monkeys . . . the preliminary

Figure 2.10. Growth in body weight of the male and female chimpanzee (from Laird, 1967).

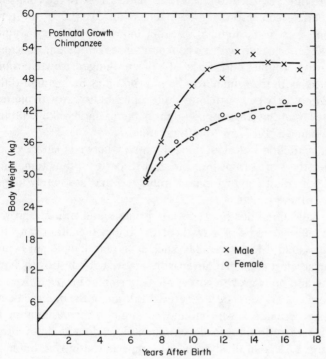

Figure 2.11. Weight velocity curve for the male and female chimpanzee (from Tanner, 1962).

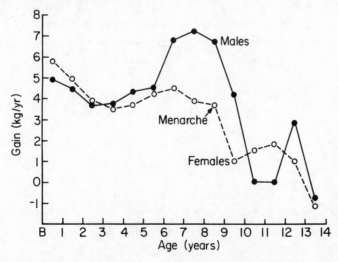

growth occupying about 1/3 and the adolescent growth about 2/3 of the time required to reach fully mature size. In the chimpanzee, adolescent growth is deferred to the last 1/2 of the total period . . . In the human a further delay has occurred so that adolescent growth with its concomitant development of sexual maturity occupies only the last 1/3 of a prolonged growth period. The delay in the human can be interpreted as being due to the *insertion,* between birth and adolescence, of two growth phases, rather than the single phase identifiable in the monkey and the chimpanzee (1967, pp. 351–2, my italic).

Humans add childhood as a period of growth between infancy and juvenescence (the start of the juvenile growth period), thus remaining dependent upon adult caregiving for a relatively and absolutely longer time than any other primate.

The other key difference between non-primate and primate growth involves the relative rates of growth of the body, the brain, and the reproductive system. Most mammals, such as the rat depicted in Figure 2.12, show an advancement of brain growth relative to body growth. Note that the brain and body growth curves are very similar in shape. Reproductive maturation of the rat occurs before the brain or the body achieves final adult size. Primates, from monkey to human, increasingly delay body growth and reproductive development, but do not delay brain growth. Figure 2.13 is an illustration of these relationships in humans. The weight of the brain reaches 90 per cent of adult size by age six and virtually 100 per cent of adult size by age 12, yet body growth continues to age 18 and beyond (note that brain growth is nearly finished before

Figure 2.12. Growth of different types of tissue in the rat (after Timiras, 1972).

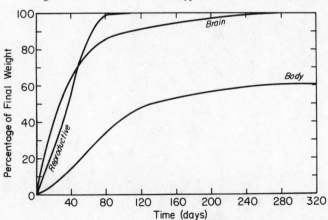

reproductive maturity even begins). This pattern of relative growth of the brain, body, and reproductive system is found in the rhesus monkey and the chimpanzee (Donaldson, 1908; Laird, 1967). Other organs of the primate body (e.g., heart, lungs, and liver) follow the body growth curve and show an increasee in growth rate at the time of sexual maturity. Thus, the brain is most unusual in its pattern of accelerated growth in comparison to other organs and to the body as a whole.

In one obvious way, this pattern of growth relates to the fact that primates are learning creatures *par excellence*. There are other consequences of the primate pattern of growth that are not as obvious as learning. For instance, rapid brain growth, deferred body growth, and progressively delayed sexual maturation greatly enhance the quality and quantity of reproductive efficiency in primates. Not only is a given offspring endowed with great flexibility of learned behavior, but also the older but still pre-reproductive youngster can help its mother provide care for new infants. In humans, with our unique childhood and adolescent periods of growth, reproductive efficiency is further enhanced and because of this, there was strong selection at the genetic level of primate biology to develop and maintain these patterns of growth.

The unique pattern of human growth

The prolonged delay in human growth due to the evolution of a childhood period of growth is one feature of the human growth curve that distinguishes it from all others. The deferred onset of sexual maturation, which does not begin until the last third of the growth period, shortens the

Figure 2.13. Growth of different types of tissue in the human being (from Scammon, 1930).

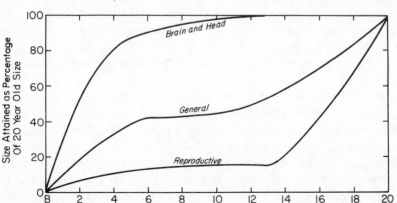

time available for the achievement of adult body size. Though it might be argued that we could take more time to grow, this was not possible for our human and hominid ancestors. Mann (1968; 1975) and McKinley(1971) analyzed the South and East African *Australopithecus* remains and found that fewer than 50 per cent of these extinct hominids lived past 20 years of age and only 15 per cent lived to age 30. Sacher (1975) estimated that *A. africanus* reached sexual maturity at 10 years (an average age for males and females). Thus, as Washburn (1981) pointed out, there was relatively little time between the end of childhood and death for a slow process of adolescent growth to occur. The evolutionary result of the competing selective pressures for a prolonged childhood period of growth, learning, and socialization versus sufficient time during adulthood for reproduction and parental care of offspring is the well-marked phase of rapid growth at the end of childhood known as the adolescent growth spurt.

The spurt is easily visualized from Figure 2.4, and is a regular feature of the growth of all human children. Only chronic and severe illness, malnutrition, or physiological stress can obliterate the growth spurt. For instance, Quechua Indian boys and girls living at high altitude in the Peruvian Andes have a late and poorly defined adolescent growth spurt. The reason may be that these children suffer from the combined stress of hypoxia (insufficient delivery of oxygen to the tissues of the body), energy malnutrition, heavy workload and cold. Their adolescent growth period is prolonged, lasting until age 22 and beyond (Frisancho, 1977).

There is some controversy as to whether the adolescent growth spurt is a uniquely human feature. Bertalanffy (1960) and Laird (1967) believe that primate, including human, adolescent growth is homologous with the postnatal growth of non-primate mammals. That is, the unique features of primate growth are the prolongation of the juvenile period, found in all higher primates, and the evolution of the childhood period, found only in humans. Others (Tanner, 1962; Watts & Gavan, 1982) believe that all higher primates share an adolescent growth period, with its characteristic spurt in growth velocity. It will be argued here, on the basis of empirical observations and evolutionary considerations, that the human adolescent growth spurt is a unique characteristic.

Watts (1985) reviewed studies of the growth of many primate species and concludes that old world monkeys, the apes, and, possibly, new world monkeys may experience adolescent growth spurts. Her evidence to support this conclusion consists of observed growth patterns that deviate from predicted mathematical curves of growth around the time of sexual maturation. Such deviations were noted in the figures of rhesus monkey and chimpanzee growth presented above (Figures 2.9 and 2.10). Most likely, these deviations are due to the onset of adult levels of gonadal hormone secretions. It is now well established that testosterone in male primates and estrogens, such as estrodiol, in female primates can increase growth rates (Martin *et al.*, 1977; Bercu *et al.*, 1983; Prader, 1984).

Some years earlier Tanner (1962), citing the work of Gavan (1953), equated these deviations from the predicted curve with the human adolescent growth spurt. The popularity of Tanner's book for medical and anthropological teaching for nearly two decades no doubt helped ingrain the belief that adolescent spurts are a primate characteristic. Gavan (1982), however, stated that his 1953 statement about chimpanzee growth was an assumption based on the opinion of Brody (1945). Brody found that no mammal, except for humans, had a postnatal growth spurt, and by extrapolation he assumed that such spurts were a primate characteristic. Gavan accepted this assumption in 1953, but in 1971, after using the computer to reanalyze his chimpanzee data, found that no spurt in linear growth could be detected. 'After all, a smoothly decelerating curve gave the best fit to most of my data with a very small residual variance' (Gavan, 1982, p. 3).

Watts & Gavan (1982) reanalyzed the same chimpanzee data and new growth data for the rhesus monkey. The chimpanzee data (growth records of nine males and seven females) were originally collected at the Yerkes Primate Laboratories between 1939 and 1954. Much of this information was used by Gavan for his 1953 paper. The rhesus monkey data were collected by Gavan between 1960 and 1967 from a laboratory colony. Longitudinal assessments of growth were made at least once a year, from birth to five years for the rhesus monkey and to age 12 in the chimpanzee.

Watts & Gavan found that simple plots of height or weight for age did not reveal a growth spurt in either the rhesus monkey or the chimpanzee. This stands in contrast to the human case, where simple graphical methods of analysis reveal the adolescent growth spurt in virtually all children. To search for non-human primate growth spurts, they employed a more 'sophisticated' method, namely exponential regression, in the form $Y = A + BR^x$. 'This model depicts growth as a process that is gradually and constantly decelerating as size increases. Its use, therefore, assumes that an adolescent spurt does not exist' (Watts & Gavan, 1982, p. 56). This is the same model Gavan (1971) used when he concluded that the chimpanzee had no growth spurt in trunk length or other linear dimensions. Watts & Gavan used this single model to fit the growth of limb segments of male and female rhesus monkeys and chimpanzees from birth through sexual maturity. They found that deviations in growth from the predicted curve followed a consistent pattern in all animals. An example is depicted in Figure 2.14 for the growth of the thigh of one chimpanzee. Deviations are positive, above the curve, in early infancy, negative in later infancy and just prior to adolescence, and then positive again during adolescence. The authors noted that the differences between the observed and predicted values at each age are small, less than a centimeter and often only a few millimeters. It was impossible for them to depict these differences graphically as two distinct curves. 'Therefore, to exaggerate the differences for drawing this figure [Figure 2.14] the actual deviations were multiplied by a factor of three' (Watts & Gavan, 1982, p. 58).

Watts & Gavan emphasized their consistent finding of growth deviations from the predicted curve as evidence for the presence of adolescent growth spurts in non-human primates. No statistical test of the mathematical significance of the deviations was made, even though Watts (1986, p. 56) states that 'the magnitude of the [adolescent growth] change is *very small*' (my italic). As stated earlier, these growth deviations at the time of sexual maturity were noted by Laird (1967) and are probably due to the increased production or release of gonadal hormones. However, the presence of these small growth velocity changes at the time of non-human primate puberty does not mean that they are equivalent to the large, and visually detectable, growth spurt humans experience during adolescence.

Coelho (1985) measured gains in crown–rump length and weight in a mixed-longitudinal sample of 250 male and 452 female olive baboons

Figure 2.14. Observed and predicted growth curves for thigh length in a male chimpanzee (from Watts & Gavan, 1982).

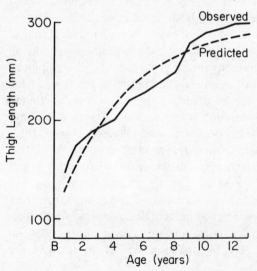

(*Papio cynocephalus anubis*). The animals were part of a laboratory colony living under naturalistic conditions in terms of the physical environment and social group composition. All animals were healthy and well nourished, and none showed signs of clinical obesity. The date of birth for all animals was known and the animals were measured once a year. The mixed-longitudinal design of the study provided data on growth between birth and eight years of age, which for this species is the total span of the growing years. Distance and velocity curves for crown–rump length and body weight are presented in Figures 2.15 and 2.16. These figures present the data in a cross-sectional fashion, which would tend to reduce the apparent size of any growth spurt (Boas, 1892). Even so, the small gains in length during puberty (puberty occurs around four years of age in this species), and the virtually constant decrease in length velocity during the entire growth period, are evidence against the belief that adolescent growth spurts occur in this monkey. Equally small increments in linear growth during adolescence were found in two other studies of monkey species, *Macaca nemestrina* and *Papio cynocephalus* (Orlosky, 1982; Sirianni *et al.*, 1982).

In contrast to relatively small changes in velocity for long bone growth, male macaques, baboons, and chimpanzees have a relatively large increase in weight growth velocity during adolescence (Figures 2.16 and 2.11). The spurt is clearly visible despite the cross-sectional method of analysis. The acceleration in weight growth is the result of an increase in size of several tissues, particularly muscle mass. Behavioral studies of

Figure 2.15. Distance and velocity curves for crown–rump length in the baboon (from Coelho, 1985).

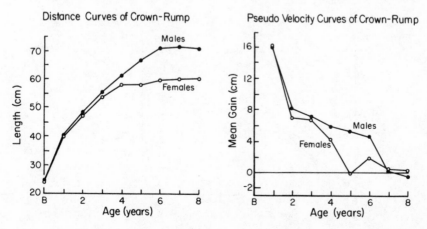

free-ranging animals show that adult males use this muscle mass for reproductive competition with each other. Female chimpanzees, baboons, and macaques have no visually detectable spurt in either linear dimensions or in weight. The absence of a growth spurt allows females to maintain a smaller adult body size compared with males. This is advantageous during pregnancy and lactation, since smaller bodies require less energy and other nutrients, allowing mothers to divert more of the available nutrients to their developing offspring. Smaller females show less of the overt aggressive sexual competition displayed by males, though female primates are no less competitive in more subtle ways, including maneuvering for breeding access to dominant males and for food (Hrdy, 1981). In this context, one may begin to understand how evolutionary pressures have shaped the type and intensity of growth spurts that non-human and human primates experience. The evolutionary pressures that shaped human growth at adolescence deserve more detailed treatment than can be provided at this point. The following chapter is devoted to this problem.

The accelerations in the growth of muscle mass or bone length that occur during the puberty of non-human primates are correlated, of course, with the endocrine changes that occur during reproductive maturation. Growth at human puberty is due to similar endocrine changes (Prader, 1984), but the human adolescent spurt in the rate of skeletal growth and the rate of growth in weight are large compared to the small change in the rate of growth of the skeleton of non-human primates. Here, then, is another qualitative difference between human and non-human adolescent growth.

Figure 2.16. Distance and velocity curves for body weight in the baboon (from Coelho, 1985).

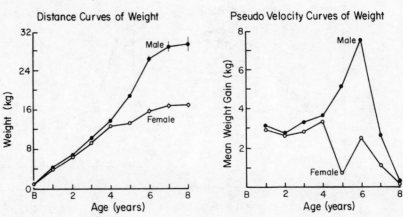

The level of hormone production of the ape versus the human cannot account fully for this difference in growth rate. In the male chimpanzee, the concentration of testosterone in blood serum prior to puberty (from one to six years of age) averages 13 nanograms per deciliter (ng/dl) (Martin *et al.*, 1977). For the human male, the prepubertal serum testosterone concentration (from ages one to 12 years) averages 9 ng/dl (Winter, 1978). The peak velocity in long bone growth of the eight male chimpanzees studied by Watts & Gavan (1982) occurred at a mean age of 10.96 years, with a standard deviation of 1.31 years. At this age, serum testosterone averages about 400 ng/dl (Martin *et al.*, 1977). Peak height velocity in human boys from western Europe occurs at a mean age of 14.06 years, with a standard deviation of 0.92 years (Marshall, 1978), when serum testosterone levels average about 340 ng/dl (Winter, 1978). Thus, the serum testosterone concentration of the chimpanzee increases about 31 fold from the prepubertal to pubertal state. In the human male, serum testosterone concentration increases about 38 fold, or 1.23 times the increase for the chimpanzee.

According to Watts & Gavan (1982), chimpanzees have a relatively small increase in the velocity of growth of individual long bones during puberty, 'usually less than a centimeter' (p. 58). In contrast, Cameron *et al.* (1982) performed a longitudinal analysis of the growth of individual limb segments in British boys. It was found that the peak value in velocity during the adolescent growth spurt ranged between 1.34 cm/yr for the forearm and 2.44 cm/yr for the tibia. From these findings one may propose that there are differences in the effect of testosterone on the skeletal growth of chimpanzees and human beings. The growth response of the human skeleton to rising testosterone levels is greater than that of the chimpanzee skeleton, since the change in the serum hormone levels of the two primates differs by a factor of about 1.23, but the change in the velocity of growth differs by a factor of at least 2.0 and perhaps as much as 4.0 or greater.

The velocity of growth in weight of the male chimpanzee is better correlated with endocrine changes at puberty, as may be seen in Figure 2.11. This observation suggests that there are differences in the growth response of skeletal and non-skeletal tissue (e.g., muscle and adipose) in the male chimpanzee. Endocrine changes occurring in female chimpanzees at puberty have not been well documented, but it is clear that they have very small increments in growth rates for both length (Watts, 1986) and weight (see Figure 2.11). It is very important to contrast this to the human case; girls experience significant growth spurts in both height (see Figure 2.4) and weight. Thus, the growth effects of pubertal endocrine changes differ between individuals of the same sex, but different species (e.g., male chimpanzees and human boys), and between the sexes within the same species (e.g., male and female chimpanzees).

Three major differences between human and non-human primate growth may be highlighted at this point. These are: (1) the residual growth potential of the non-human versus the human primate at adolescence, (2) the sensitivity of different body tissues to growth promoting stimuli, and (3) sex differences in the expression of growth spurts at adolescence. The monkey, ape, and human all experience a delay in sexual maturation and a prolongation in growth. In the human the delay is both relatively and absolutely greater than in the monkey or ape (Laird, 1967). In addition to this, humans also have a markedly increased potential for growth in height and weight during adolescence. This growth potential is likely to be regulated more by the sensitivity of neuroendocrine receptors and post-receptors (i.e., biological tissues) to growth stimuli than by the rate or amount of production of the stimuli (e.g., hormones) themselves. The lack of linear associations between testosterone concentrations and growth velocities in skeletal and non-skeletal tissue of chimpanzees and humans shows this. The differences in cellular sensitivity to growth stimuli between non-human and human primate growth are probably controlled at the genetic level, but are not the result of the evolution of new structural genes. Rather, control lies in the regulatory genes that initiate and terminate each of the distinct periods of growth and control their duration (Britten & Davidson, 1969; King & Wilson, 1975).

The philosophy of human growth

Non-human primate models are used to study human growth because of similarities in anatomy and physiology between the species. Also, there is an assumed evolutionary continuum between the living primate species – *Macaca mulatta, Pan troglodytes,* and *Homo sapiens* (the rhesus monkey, the chimpanzee, and the human being). There is an evolutionary connection relating these species, however, the living monkeys, apes, and humans each have a separate evolutionary history. Cercopithecines (Old World monkeys) and hominoids (apes) separated some 20 million years ago and the hominoid–hominid (ape–human) split occurred at least six million years ago. There is no evolutionary reason to expect that the patterns of growth of these three divergent and ecologically distinct species should be identical or even similar. As Gavan (1971, p. 54) observed, chimpanzee postnatal growth begins 'with an initially high rate which decelerates smoothly as size increases, but it is well known that human growth is characterized by a growth spurt . . . Some change must have occurred in human growth since we and the chimpanzee have had a common ancestor.'

The notion of an evolutionary continuum, a 'great chain of being' (Lovejoy, 1936), is a popular cultural construct in Western society. In its original usage it implied, erroneously, that all living creatures, from amoeba to human, form a living evolutionary sequence from the simplest to the most complex creature. We now understand that humans are not the culmination or the goal of evolutionary history. We are just one of ore than two million animal species alive today, each the end product of its own history and each with its own unique place in nature. Yet, the 'great chain' is sometimes misapplied to the connection between human and non-human primates. Some observers tend to see the ape more as a model for human biology and behavior than as a creature in its own right.

This point was cogently argued by Scott (1967, p. 72): 'Subhuman primates are not small human beings with fur coats and (sometimes) tails. Rather they are a group which has diversified in many ways, so that they are as different from each other [and humans] as are bears, dogs and raccoons in the order Carnivora.' Though Scott referred specifically to psychological attributes of species within the orders Carnivora and Primates, his cautionary remarks apply equally to morphology, physiology, and, in the present context, patterns of growth. As mentioned above, within the Carnivora certain social species (dogs, wolves, and lions) experience a decrease in growth velocity between infancy and adulthood, which corresponds to the juvenile growth phase of primates. All other Carnivora mature from infancy to adulthood without alterations in growth rate. There is no reason to expect, *a priori*, that the Primates, as an order of mammals, would be any more uniform in growth patterns than the Carnivora. Newell-Morris & Fahrenbach (1985) reviewed the use of non-human primates as models for human development and growth and concluded 'there are problems with the extrapolation from the nonhuman primate model to the human condition because of intergeneric differences in size, growth and development rates, and timing. Although investigators justify direct extrapolation of their findings on the basis of the close genetic relationships of all primates, this assumption in many cases may be little more than absolute faith in the evolutionary argument from which it stems' (p. 35).

A recent attempt to use the growth and development of non-human primates as a conceptual model for human growth is that of Gould (1977). Arguing from an idea originally proposed by Bolk (1926), Gould suggests that the major difference between human and non-human primate growth is that humans mature sexually while still in an infantile or child-like stage of physical development. Compared with the chimpanzee we are neotenous, or fetalized, in our body growth and in our physical features. The infant chimpanzee (Figure 2.17) exhibits many features of the adult human: a large rounded cranium, flat face, and erect posture. The adult chimpanzee exhibits a prognathic face due to hypermorphic growth of

Figure 2.17. Profile sketches of an infant and adult chimpanzee (drawn from photographs in Gould, 1981).

jaws and semi-erect posture due, in turn, to its knuckle-walking and brachiating modes of locomotion.

Actually, Gould presented two levels of argument in favor of neoteny as the pattern for human evolution and growth. The first level may be called his popular argument and the second level his scientific argument. In his popular argument he stated that the pattern of human growth retains infantile features of our hominoid ancestors into present-day human adulthood. 'In neoteny rates of development slow down and juvenile stages of ancestors become adult features of descendants. Many central features of our anatomy link us with the fetal and juvenile stages of primates' (1981, p. 333). On the same page Gould stated 'If humans evolved, as I believe, by neoteny, . . . then we are, in a more than metaphorical sense, permanent children.' Gould's scientific notion of neoteny was presented in an earlier work (1977). There he rejected the notion that adult humans represent the fetalized, or juvenilized, stages of living or ancestral primates. Instead, he argued that adult humans are paedomorphic, that is, retaining as adults the characteristics of the fetal, infantile, and childhood periods of development. In a vague way, neoteny was retained in the scientific argument as the mechanism for human development. In Gould's phrase, 'Our paedomorphic morphology is a consequence of retarded development; in this sense, we are neotenous' (1977, p. 397).

The concepts of paedomorphism and neoteny may serve as metaphors for the evolution of the human growth pattern that is described in this chapter. However, attempting to convert these metaphors into the biological processes that underlie human development is like attempts to equate the workings of electronic computers with the neuroendocrine basis for the operation of the human brain. As Gould pointed out, one aspect of human development is a delay, or retardation, in growth and sexual maturation. However, this delay is not merely an extension of a general primate pattern that simply carried childhood features into adulthood. Kummer (1953), Bertalanffy (1960), and Starck & Kummer (1962) argued on the basis of craniofacial and postcranial growth that 'the concept of "fetalization" is to be refused with respect to the ontogeny and evolution of the human . . . for this is not the result of an arrest of growth at an early phase but of a differentiation in changed direction . . . Hence we may speak of "retardation" but not of fetalization in human development' (Bertalanffy, 1960, p. 250).

One specific example of this view of human development is the ontogeny of cranial growth related to language development. The human newborn cannot produce the speech sounds (phonemes) used by adult speakers of any language. Lieberman *et al.* (1972) and Laitman &

Heimbuch (1982) believed that the shape of the basi-cranium is the reason for this. They argued that newborns possess a basi-cranium with a relatively large angle of flexion (the angle formed by the junction of the occipital and vomer bones). This angle influences the shape of the soft tissues of the vocal track, especially the pharynx, and determines the nature of the vocal sounds the newborn can produce. During growth the angle of flexion becomes more acute, as the skull assumes child, juvenile, and adult proportions, and linguistically recognizable phonemes are producible. In this one aspect of growth of the skull and its functional correlates, the metaphors of neoteny and paedomorphism are clearly not useful.

In most other aspects of growth the neoteny metaphor is equally useless. The human pattern of growth differs from other primates in at least two ways: the addition of the childhood period and the markedly increased potential for statural growth at adolescence. These are new features of the human growth curve. A graphic case against the paedomorphic metaphor for human development was presented in Figure 1.3. From fetus to child to adult, body proportions are so much altered that the mature morphology cannot be simply predicted from earlier stages of growth. Allometry, differential rates of growth of parts of the body relative to that of the body as a whole (Huxley, 1932), is the rule in primate development. Both positive and negative allometry take place in the ontogeny of human development (e.g., leg versus trunk growth, or head versus body growth). These allometric changes bring about functional differences between the adult and child in physical appearance and performance.

Ontogeny from childhood to adulthood entails the development of new biological competencies, especially those related to reproduction and intelligence. The onset of reproductive maturation is best represented physiologically and mathematically as a catastrophic alteration of the body's neuroendocrine system; it is unlike anything operative in the prepubertal child (Bogin, 1980). The stage theory of Piaget (1954; Piaget & Inhelder, 1969) provided a descriptive and theoretical understanding of the development of human intelligence. Though the neoteny metaphor correctly asserts that adult humans possess an intellectual plasticity and curiosity usually found only in the young of other species, Piaget's stage theory of human development showed that this is due to the maturation of increasingly sophisticated and flexible cognitive processes, rather than due to the retention of childhood intellectual abilities. At the physiological level, Eccles (1979) showed that the adult human potentials for playfulness, creativity, and intellectual advancement are not derived paedomorphically, rather they are new competencies derived from a constant remodeling, restructuring, and maturation of the neurological architecture in the central nervous system.

A philosophy of human growth and development that emphasizes the progressive appearance of new biological and behavioral traits is more satisfying empirically, and intellectually, than a view of development that emphasizes growth retardation and the permanency of childhood. An acceptable philosophy must also acknowledge the mammalian and primate origins of the human pattern of growth. Combining each of these ideas into an holistic conceptualization of growth allows one to understand the quantitative differences in development between species and the qualitative uniqueness of the human growth pattern.

3 *Evolution, ecology, and human growth*

The pattern of human growth is characterized by a prolonged period of infant dependency, an extended childhood, and a rapid and large acceleration in growth velocity at adolescence leading to physical and sexual maturation. These traits are considered to be advantageous for our species since they provide:

(1) an extended period for brain development,
(2) time for the acquisition of technical skills, e.g., tool making and food processing, and
(3) time for socialization, play, and the development of complex social roles and cultural behavior.

These statements are standard 'textbook' rationalizations for the value of the pattern of human growth, but they do not explain how that pattern of growth evolved, that is, they do not provide a causal mechanism for the evolution of human growth. Rather, these are tautological statements, arguing for the benefits of the simultaneous possession of brains that are large relative to body size, complex technology, and cultural behavior. First causes may not be deduced from this type of circular reasoning, but the big brain–technology–culture argument is uncritically accepted by many students of human evolution. For instance, after reviewing some of the evidence for human evolution one primatologist states that 'human evolution is a paradox. We have become larger, with long life and immaturity, and few, much loved offspring, and yet we are more, not less adaptable.' The alleged paradox is resolved by concluding in the next sentence that 'mental agility buffers environmental change and has replaced reproductive agility' (Jolly, 1985, p. 44). By this it is meant that we are a reproductively frugal species compared with those that lavish dozens, hundreds, or thousands of offspring on each brood or litter. However, it is the emphasis on brains rather than reproduction that is paradoxical. Evolutionary success is traditionally measured in terms of the number of offspring that survive and reproduce. Biological and behavioral traits do not evolve unless they confer upon their owners some degree of reproductive advantage, in terms of survivors a generation or more later.

In light of this, additional reasons must be added to the list of benefits

74

of the human pattern of growth. Employing an evolutionary and ecological perspective, this chapter will present arguments for the inclusion of the following four additional reasons for the human growth pattern:

(4) Slow childhood growth and delayed maturation may have originally evolved as feeding adaptations. The addition of fibrous, 'tough-to-chew' foods in the diets of our ancestors may have selected for a delay in molar tooth eruption which allowed for efficient processing of these foods throughout the life cycle. Delays in dental eruption produced, as an indirect byproduct, delays in growth and maturation in other systems of the body.

(5) Small body size prior to adolescence, coupled with delayed maturation, tends to reduce intra-family and intra-group hostility. Small body size of juvenile young reduces competition with adults for food resources, because slow growing, small juveniles require less total food than bigger adults. Delayed maturation is also advantageous since it eliminates some social and sexual competition between adults and young during the years when technological, social, and cultural learning occur.

(6) Small body size of juvenile young also reduces competition for food resources with infants and young children. Since adults are compelled to feed infants and young dependent children, feeding competition with these younger offspring is minimized so as not to bring the interests of juveniles into conflict with the interests of adults.

(7) The human growth pattern of relatively early neurological maturity versus late sexual maturity allows juveniles to provide much of their own care and also provide care for young children. This frees adults for subsistence activity, adult social behaviors, and further childbearing. The economic, social, and reproductive value of juveniles, as 'babysitters,' gives delayed sexual maturation added selective advantages.

The justification for adding these four reasons for the evolution of the human pattern of growth requires a review of some of the fossil and paleoecological evidence for human evolution. The review of these fossils, and their ecological setting, focuses on the evidence for the evolution of the human growth pattern.

Early hominid paleontology and ecology

The paleontological evidence indicates that some of the essential features of human growth appeared at an early time in our history. The genus *Australopithecus*, which appears to date from about 3.7 million years ago, followed a pattern of growth that included a delayed dental eruption sequence, relatively rapid growth of the central nervous system during infancy and childhood, bipedalism, and, possibly, a childhood growth period and delayed sexual maturation. It is likely that a hominid species (hominid species include living humans and fossil species that are ancestral to living humans) possessing this growth pattern was more reproductively successful than species without it. However, all of these features could not have appeared simultaneously in any early hominid. There are too many separate structural and functional changes in dental, skeletal, and, by implication, physiological and muscular systems for a single evolutionary event to be responsible for the sudden appearance of each of the separate traits in one organism. The evolution of human growth must have begun some time before the appearance of *Australopithecus*.

To account for the evolution of hominid and human growth, an ecological perspective must be taken. We may understand the evolution of human growth only if we can define the constraints of the ecological setting in which this evolution occurred. One part of ecology is concerned with the availability and distribution of food resources in the environment. Paleontologists often focus their attention on feeding ecology when developing explanatory models of human evolution. A classic example of this is the 'seed eater' hypothesis proposed by Jolly (1970). Jolly's model accounts for the evolution of hominid physical characteristics such as non-projecting canine teeth, enlarged molars, rotary chewing, opposable thumbs, and bipedalism as a consequence of our ancestors' adaptation to a new diet of seeds and other foods that come in small, tough-to-chew packages. A series of growth changes was required to achieve each of these 'seed eater' anatomical features. For instance, molar enlargement involved the growth of not only more dental tissue, but also the supporting bone in the mandible, maxilla, face, and cranium. Bipedalism required proportionally greater growth of the legs compared with the arms, and structural growth changes in the shape of bones from the cranium to the toes (Campbell, 1985).

Jolly links together two lines of evidence to develop his feeding ecology model of hominid evolution. The first line of evidence is that our ancestors were among a group of hominoids (a group that includes humans, apes, and their ancestors) that began to exploit a new habitat, the tropical woodlands and savannas. Jolly points out that a world-wide decrease in rainfall during the Miocene period (25 to 6 million years ago) resulted in the shrinking of the tropical forests that were the home of early hominoids, and the expansion of the woodland–savanna zones. Jolly, and more recently Pilbeam (1979; 1984), propose that the fossil evidence indicates that Miocene hominoid primates underwent an adaptive radiation in speciation and biological variation as they began to exploit the new habitat zones.

Jolly reasoned that, in these new habitats, hominoid primates – ex-forest dwellers who were essentially vegetarians – would have adapted to new plant foods, rather than assume a completely different diet such as carnivory. Plants adapted to dry or seasonally dry environments tend to store most of their nutrients in seeds, roots, and tubers; such tissues are more resistant to drought and heat than leaves or stems. More recently, Walker (1981) and Kay (1985) studied the finer details of early hominid dental structure and tooth wear using the scanning electron microscope and tooth wear experiments. These researchers propose that the diet of the early hominids, including *Australopithecus* spp., was largely herbivorous, but included softer plant foods (leaves and fruits) as well as seeds and tubers. Although the cuisine may not have been as catholic as Jolly proposed, the new foods in the early hominid diet were the tough seeds and tubers. Eating them may have required the anatomical changes listed above, for instance, precisely opposable thumbs to grasp small seeds, short canines to allow rotary grinding, and large molars to pulverize effectively the tough cellulose coating surrounding the nutrients these foods contain. Bipedalism may have been a consequence of the evolution of hands dedicated to feeding rather than locomotion. The implications for hominoid growth patterns were that amounts and rates of growth of the jaw, face, and skull changed as a result of molar enlargement, and that growth of the extremities changed as a response to bipedalism. For instance, legs grew relatively longer than arms, pelvic bones shortened and broadened, and feet assumed a shape dedicated to support and locomotion rather than manipulation.

Similarities in anatomy and behavior between humans and the gelada baboon (*Theropithecus gelada*) were used by Jolly to support his speculations on paleoecology and hominid evolution. The gelada lives in a semi-arid savanna habitat where grasses, seeds, and roots are the predominant foods. They have evolved a likeness to humans in order to gather and process these foods. The gelada likeness to humans includes more precisely opposable thumbs, broader molars with flatter crowns, shorter faces, shorter canines, greater ability for rotary chewing, and more trunkal uprightness compared with other baboons adapted to wetter environments.

There is a considerable body of fossil evidence in support of the 'seed eater' hypothesis. A group of extinct primate species, including *Ramapithecus, Gigantopithecus,* and others, collectively called ramapithecines by Pilbeam (1984), exhibit many of the traits predicted by the Jolly hypothesis. Some of these ramapithecine fossils also exhibit the beginnings of the hominid pattern of delayed dental eruption. These fossils are from the Miocene period, dating from about 14 to 8 million years ago, which is the predicted time period for the change in feeding adaptation. The ramapithecines were hominoid primates, not hominids, and their evolutionary affinities may be closer to the living apes, especially the orangutan, than to the hominids. However, pre-*Australopithecus* hominids have not yet been identified unequivocally. Thus, a description of ramapithecine dental characteristics provides a useful analogy for the evolution of hominid dental traits.

Pilbeam states that the ramapithecine fossils have robust mandibles, with evidence for the attachment of powerful chewing muscles, and their molar teeth have thick enamel on the occlusal (chewing) surfaces. These features distinguish them from other contemporary fossil hominoids (e.g., *Proconsul* and *Dryopithecus*) and were the result of new patterns of dental and skeletal growth. Ramapithecine masticatory traits were likely to have been associated with the demands of the diet of these creatures. Paleontologists such as Pilbeam (1984) and Wolpoff (1980) point out that it is difficult to make detailed reconstructions of the paleoecology of the rampithecines. However, using the available evidence, Tattersal (1975) proposed that they lived in open woodland and savanna habitats, exactly the places predicted by the Jolly hypothesis. Tattersal and Pilbeam believe that the new patterns of dental and facial growth, and the masticatory apparatus of the ramapithecines, were likely to have been adaptations to the grinding and chewing needed to extract nutrients from seeds, tubers, and roots.

The molar teeth of ramapithecine specimens from Asia (*Ramapithecus* and *Gigantopithecus*), from Southern Central Europe (*Rudapithecus, Bodvapithecus,* and *Ouranopithecus*), and the Near East (*Ramapithecus*) display a pattern of uneven wear, which Simons & Pilbeam (1972) and Wolpoff (1980) attribute to a delay in dental growth and maturation. These paleontologists find that in adult speciments of ramapithecines with all three permanent molar teeth present, the first molar is often very worn, sometimes the enamel on the occlusal surface is almost completely obliterated, the second molar shows less wear than the first molar, but in places the dentin is exposed, and the third molar is little worn, sometimes it is virtually pristine in appearance. Simons & Pilbeam state that the differences in wear between the first, second, and third molars in these fossils are greater than those seen in the living apes (gorillas or chimpanzees) of known ages and similar to living humans who consume traditional, largely unprocessed, foods.

This wear occurs despite the fact that ramapithecines, and humans, possess molar teeth with thicker enamel on the occlusal surface than those of the African apes. In people this wear pattern results from the delay in the rate of growth of the permanent dentition. For instance, the first molar of chimpanzees erupts at about 3.5 years, the second at 7.0 years, and the third at about 10.5 years. In humans the delay in the rate of dental growth results in the eruption of these same teeth at approximately 6.5, 11.5, and 16.5 years. Thus chimp molar teeth grow and erupt about 3.5 years apart, and human molars grow and erupt about 5.0 years apart. The total time span of the growth period from eruption of the first to the third molar is 7.0 years in the chimp and 10.0 years in the human. These are only average figures; some humans take 12 or more years to complete the eruption sequence.

For living humans the time span from eruption of first to second molar, roughly 6 to 12 years of age, approximately corresponds to the length of the childhood growth period. The time span between the eruption of the second and third molars approximately corresponds to the adolescent growth period. It is unlikely that these associations may be applied directly to the ramapithecines. However, their delay in molar eruption may indicate a delay in total body growth and maturation compared with species of living apes. Such extrapolation seems justifiable, for in order to evolve delayed molar growth and eruption, which helps to assure that at least one good pair of occluding molars will be available for chewing at all times, the total growth rate of the body must be delayed as well. The biological control of dental and skeletal growth is regulated by different systems in the body, as demonstrated by the need to use separate standards to assess the maturational status of children from their teeth or bones (Demirjian, 1986). Despite their biological independence a correlation exists between dental and craniofacial growth (Sullivan, 1986). Compared with other primates, human jaw and facial growth during childhood and adolescence is delayed so as to be in concert with tooth eruption and to maintain the functional integrity of the masticatory complex. Since the same general body systems (hormonal and neurological) that regulate skeletal growth and maturation in the face and head also regulate post-cranial skeletal growth, delayed molar growth and eruption is associated with a total delay in growth and maturation of the body. This is the case for living humans and may well have been the case for the ramapithecines.

The initiation of the human pattern of growth, especially delayed molar eruption and an extended period of growth, may be related to a feeding adaptation made by our ancestors during the Miocene geological period. Processing new foods may have initiated the evolution of the human growth pattern, but other selective pressures were required to continue and direct this evolution to its present-day outcome. With the appearance in the fossil record of the next major hominid group, the australopithecines, we may derive an indication of what these selective pressures may have been.

The earliest species of *Australopithecus, A. afarensis,* dated to about 3.5 million years ago, shows evidence for a delayed molar eruption sequence, which is more strongly developed than in the ramapithecines. In addition, *A. afarensis* possessed the uniquely hominid traits of bipedality and hands designed for manipulation of objects rather than locomotion. New patterns of growth brought about the development of bipedality and forelimb manipulation in the australopithecines. Issac (1978) proposed that bipedality evolved as a feeding adaptation, designed to allow for the collection of different foods from a variety of sites in the woodland–savanna environment and for the transport of these foods, by hand carrying, to places for processing, sharing, eating, and, possibly storage. Tanner & Zihlman (1976) and Lancaster & Whitten (1985) add to this list of possible causes the need for early hominid females to carry their infants. Women living in hunting and gathering societies, and in many horticultural and agricultural societies, carry their nursing infants with them all day. Presumably, so did the lactating mothers of our early hominid ancestors. The reasons why primate infants must remain in close proximity to their mothers and be fed virtually on demand were presented in Chapter 2. Human infants have neither the physical ability to cling to their mothers nor the maternal body hair to which to cling. Thus, women must carry their infants. Newman (1970) and Frisancho (1979) speculate that our early hominid ancestors possessed a distribution of body hair similar to our own. If this was so, then australopithecine mothers would have been obliged to carry their infants, rather than have the infants cling to their bodies.

The cause of bipedalism may be unclear, but its consequences for changes in growth rates throughout the musculo-skeletal system are clear. The changes include structural alterations in the foot for support and gait, lengthening of the legs relative to the arms, shape changes in the pelvis and the spine, a repositioning of the skull so that it is more centrally located above the cervical vertebrae, and a repositioning of muscular origins and insertions throughout the skeleton (Campbell, 1985). These structural alterations are mostly achieved by different rates of growth of various parts of the body during infancy, childhood, and adolescence. This can be seen clearly in relation to bipedalism. During infancy, when the child is dependent on the mother for care and feeding, the rate of growth of the legs and the maturation of corresponding nerve cells in the motor–sensory cortex of the brain are slow relative to the growth of the head, the arms, and the nerve cells that control movement in the upper half of the body. By about two years of age, there is an acceleration in the rate of leg growth and maturation of the motor–sensory cortex region devoted to the legs (Tanner, 1978). At the time the child is weaned completely from the breast (about four years of age in many pre-industrial societies), a time when independent locomotion will take on greater importance for the child, the rate of leg growth is faster than the rate of arm growth (Scammon & Calkins, 1929; Hansman, 1970).

In the hominid fossil record, *A. afarensis* is followed by the appearance of at least two new species, *A. africanus* and *A. robustus*. Anatomically, these species possessed two new dental characteristics that may have helped maintain selective pressures for the continuing evolution of the human growth pattern. The new dental characteristics are canine teeth that are relatively shorter than those possessed by *A. afarensis*, and premolar teeth that are molar-like in crown shape. One interpretation of these dental changes is that the cutting and chopping functions of these teeth were no longer needed. Instead of teeth, *A. africanus* and *A. robustus* may have used stone tools for some food preparation (Bartholomew & Birdsell, 1953; Wolpoff, 1980). Issac (1978) and Potts (1984) interpret evidence from East African *Australopithecus* archaeological sites to indicate that by 2.0 million years ago some hominids were habitual makers and users of stone tools. Of course, the use of tools made from parts of plants and animals or unmodified rocks probably preceded the use of flaked stone tools. The use of 'natural' tools would have contributed to the selection pressures shaping the course of human evolution and may have been a co-determinant, along with the 'seed eater' selection pressures, of hominid biology. Issac and Potts also propose that the stone tool making early hominids lived in nomadic social groups that gathered, scavenged, and hunted food, and that at first the stone tools, and later the foods, were often collected, prepared, and utilized communally. This technological and social sophistication was coupled with an increase in brain size; these hominids had brains relatively larger in proportion to body size than *A. afarensis* or any living ape.

Given this evidence for biology and behavior, Mann (1975) characterized *Australopithecus* as a genus dependent on a long childhood for the learning of technological and social skills required for adult survival. Mann's emphasis on learning as the reason for a long childhood followed from an opinion of Dobzhansky (1962) that 'Although a prolonged period of juvenile helplessness and dependency would, by itself, be disadvantageous to a species because it endangers the young and handicaps their parents, it is a help to man because the slow development provides time for learning and training, which are far more extensive in man than in any other animal' (p. 58). This view, that the evolution of prolonged childhood dependency in humans is a 'necessary evil,' is common in anthropological and biological writings. It encompasses the first three advantages of the human pattern of growth outlined at the beginning of this chapter. Its corollary argument is that since humans, and our hominid ancestors, are cultural animals and since culture is a set of learned behaviors, a long childhood is required to acquire culture.

Cultural behavior is taken to include the dependence of early hominids on stone tools (technology) and the food sharing and communal living that took place at temporary encampments (social organization) by 2.0 million years ago (Issac, 1978; Potts, 1984). Culture also includes the 'standards of appropriate behavior, learned by the individual, which allow him successfully to interact within his own social and group milieu' (Mann, 1972). At this level, culture consists of the ideologically defined rules and reasons for particular behaviors. In turn, dependence on technology, complex social organization, and ideologically defined behavior acted upon biological development in a classic feedback manner. According to Mann, the result for hominids was that general body growth and sexual maturation became progressively delayed so as to allow the young time enough to acquire culturally transmitted survival skills before they would need to use these skills to attract mates and rear offspring. The peculiar neurological development of hominids may also be explained by this feedback model; the brain grew and matured faster than other parts of the body for efficient learning of a myriad of technological facts, social relationships, and ideological constructs that make for reproductively successful adults (Hallowell, 1960).

This summarizes the traditional view of the adaptive significance of the human growth pattern. Proponents of this view uniformly cite the advantages of learning and culture, but pay virtually no attention to the reproductive and social difficulties that dependent young impose on their caregivers. Moreover, this traditional view ignores the possibility that slow rates of growth and delayed maturation may not only aid the survival of individual hominid chidren, but also may directly enhance their parents' survival and total reproductive efficiency.

Early hominid demography

It is indisputable that childhood is the time when the bulk of human cultural and social learning takes place, and that young children, younger than five or six years old, are dependent for their survival upon older caretakers. Mann's interpretation of the physical and behavioral remains of *Australopithecus* finds that they were similarly dependent during childhood. This conclusion was based on his (Mann, 1975) analysis of the fossil evidence, which indicates that the timing of dental development of *A. africanus* and *A. robustus* is essentially the same as that for living humans. The specimens of *Australopithecus* that Mann examined were those from South Africa, and although exact dating of these fossils is difficult they are believed to have lived between 3.0 and 1.5 million years ago. The stages of dental development and tooth eruption of these fossils were compared with those of living apes and humans. Apes mature more rapidly and complete their growth sooner than humans, including the eruption of the permanent molars. This means that the unerupted teeth of apes must develop at a faster rate than the unerupted teeth of humans. Mann compared X-ray photographs of juvenile australo-pithecine mandibles with the first or second permanent molar just erup-ted, with mandibles of chimpanzees and humans at the same stage of eruption. The amount of development of the unerupted second or third australopithecine molars was, in every case, like that of living humans. *Australopithecus* fossils did not have the rapid molar development of chimpanzees and were not 'half-way' between the chimpanzee and the human pattern. As explained above, a human-like delay in dental development indicates a similar delay in overall maturation of the body. From this association, Mann speculated that the disadvantages of delayed growth, such as increased risk of morbidity and mortality prior to reproductive age, must have been offset by more important advantages for this pattern of growth to evolve. The learning and culture argument outlined above was Mann's choice for the offsetting advantages.

Growth delay, whatever its cause, does not account for the evolution of the human adolescent growth spurt. After all, within the order Primates, from the prosimians to the apes, there is an increasing delay in growth and development at all stages of life (Schultz, 1969), but none of these species has an adolescent growth spurt. With the australopithecines the trend towards delayed development may have reached a critical demographic threshold. It seems that *Australopithecus* suffered from a high degree of relatively early mortality. Mann (1975) also studied the degree of occlusal wear on the molars to estimate age at death. He used rates of occlusal wear of pre-industrial human populations to make these estimates for the fossils. A total of 168 individual fossils were aged in this manner, and the average age at death was estimated to be between 17.2 and 22.2 years. McKinley (1971) calculated the age at death of the South African australopithecine specimens studied by Mann and East African australopithecine fossils. Using a mathematical demographic model of survivorship, McKinley estimated the average age at death for all the fossils studied to have been 19.8 years.

The interpretation of australopithecine growth and development offered by Mann and McKinley is based upon a comparison of tooth formation, eruption, and wear patterns between the fossil hominids and living apes and humans. Lewin (1985) stated that it may be misleading to consider 'early hominids simply as diminutive humans' or as essentially ape-like. Following this admonition, Bromage & Dean (1985) developed a new method of estimating the age at death of hominoids that is supposed to be applicable to all species, living and extinct. The method, based on the formation of micro-structural features of incisor tooth crowns, indicated that *Australopithecus* and early *Homo* may have grown at a rate more similar to apes than to humans. If so, age at death of the immature australopithecines would have to be revised downward by an average of 2.24 years for the five fossil specimens reported by Bromage & Dean. The incisor tooth aging method is dependent on many circumstantial correlations between different aspects of crown formation. Moreover, the method is speculative when applied to hominids, since there are no standards for living or extinct humans. However, using yet a third method of dental aging, Smith (1986) finds results similar to those of Bromage & Dean. Smith constructed a series of pattern profiles of the development of tooth crowns and roots for living pongids (chimpanzees and gorillas) and humans. The pattern of crown and root formation of several species of early hominids (*A. afarensis, A. africanus, A. robustus, A. boisei, Homo habilis,* and early *H. erectus*) was compared with the pongid and human pattern. None of the early hominids conforms precisely to either the pongid or the human pattern profiles. However, *A. robustus* and *A. boisei* have profiles closer to the human pattern, while all of the other fossil specimens have profiles closer to the pongid pattern. In essence, Smith concludes that early hominids had patterns of tooth formation and rates of development unlike those of any living primate.

To place australopithecine growth, development, and life expectancy in perspective, data are presented in Table 3.1 for the survivorship, age at menarche (first menstruation), and lifespan of the chimpanzee, fossil hominids, and pre-industrial human samples. Australopithecine survivorship to age five is greater than that for chimpanzees, the Libben paleo-Indians (a sample of 1327 burials dated to about AD 800–1100), and some living pre-industrial people. By age 15 differences between the hominid samples begin to diminish, but chimpanzee survivorship remains less than that for any hominid. During the early adult years, ages 20 to 25, survivorship is about equal in each of the hominid samples, but by age 30 the modern human groups show greater survival than *Australopithecus*.

The ages at menarche and the lifespan predictions for the fossil and Libben samples were estimated by Sacher (1975) using a regression formula based on brain weight and body weight. Though these two

Table 3.1. *Calculated survivorship, age at menarche, and lifespan of the chimpanzee, fossil hominid samples, and pre-industrial human populations*

| Sample | Per cent surviving to age | | | | | | Age at menarche[1] | Lifespan | |
	1	5	15	20	25	30		Predicted[2]	Observed[3]
Chimpanzee (Goodall, 1983)	71	63	53	43	35	20	10.5–11.3	43	<40
Australopithecus (McKinley, 1971)	95	81	66	47	30	15	10 (?)	47	<40
Libben paleo-Indian (Lovejoy et al., 1977)	83	69	53	46	41	35	14.5	89	<60
!Kung San (Howell, 1979)	80	66	58	56	53	51	16–17	89	88
Yanomamo (Neel & Weiss, 1975)	73	62	50	43	36	30	13–14	89	70

[1] *Australopithecus* and Libben estimated by the formula of Sacher (1975), others as reported in the reference cited under 'Sample' column.
[2] Predictions based on the formula of Sacher (1975).
[3] As reported in the references cited under 'Sample' column.

weights are also estimates for fossils, Sacher calibrated his formula against a sample of 239 species of living mammals from 12 orders including primates. For human beings, Sacher's predicted lifespan is 89 years, very close to the observed maximum lifespan of about 90 years for pre-modern societies. Only a few individuals in these societies reach the maximum age; many of those people who claim to be so old are likely to have overstated their ages (Mazess & Mathisen, 1982; Rosenwaike & Preston, 1984). In the case of the !Kung the maximum known lifespan is 88 years for one individual, but life expectancy at birth is only 35 years for the population (Howell, 1979). The chimpanzee menarche and observed lifespan data are for wild-living animals (Gombe Stream Reserve, Tanzania). Captive chimpanzees reach menarche at a mean age of 8.9 years (Goodall, 1983) and have lived to age 49 years (Sacher, 1975). The relatively late age at menarche of the !Kung may be due to the combined stress of under-nutrition and a heavy workload (Howell, 1979). The estimate of age at menarche for *Australopithecus* is the most speculative of any on this table. Menarche usually occurs after the formation and eruption of the second permanent molar tooth in both chimpanzees and people. The mean age at eruption of the chimpanzee second molar is about 6.5 to 7.0 years (Dean & Wood, 1981) and menarche occurs no earlier than at a mean age of about 8.9 years. For people the corresponding ages are 12 years for eruption (Dean & Wood, 1981) and about 13 years for menarche for girls of middle or upper socioeconomic classes (Marshall & Tanner, 1986). Using the rates of australopithecine dental development calculated by Bromage & Dean (1985) and Smith (1986) it may be reasonable to place the age of menarche for these early hominids within the chimpanzee range. Menarche is an indicator of sexual maturation, but it is not a marker of fertility. Between menarche and first conception there is a period of adolescent sterility. Copulations occur during this time but no pregnancy results. For the chimpanzee this period of adolescent sterility lasts between one to three years (Goodall, 1983); for girls the period is usually about the same (Lancaster & Lancaster, 1983).

These estimates of survivorship indicate that the cumulative mortality of australopithecine populations may have been greatest just after growth of the body was virtually complete and sexual maturity was achieved, that is, after age 15. A smaller percentage of chimpanzees live to the end of their growth period. Modern humans (the Libben, !Kung, and Yanomamo samples) experience, generally, a larger percentage of survival past the growth period. These data also show that the *average* age at death of the australopithecines cannot be taken as an indication of the *number* of individuals surviving into their third or fourth decade of life.

Obviously, this number was sufficiently large to maintain australopithe-cine survival for one million or more years.

One must make some strong assumptions to accept fully any interpreta-tion of the fossil evidence for age at death and early hominid patterns of growth and development. For instance, Brain (1981) proposed that the fossil assemblages of *Australopithecus* may not be those of biological populations, rather they may represent the activity of carnivores that selectively preyed upon young and old individuals. Moreover, there is statistical evidence that *Australopithecus* may have been longer-lived than indicated by the fossils we have. As shown in Table 3.1, Sacher estimated an upper limit of 47 years for the lifespan of *Australopithecus*, though no known fossil is believed to have lived to that age. Harvey & Zammuto (1985) showed that age at death and reproductive age were strongly correlated ($r = 0.90$ or higher) in 25 mammals of various taxonomic orders (not including primates). The strength of the correla-tion remained even after the effects of body size were removed. By this analysis, relatively delayed growth, maturation, and late reproductive age in *Australopithecus* would predict a relatively late age of death in this species. The known causes of death for chimpanzees, the Libben sample, the !Kung, and the Yanomamo show that disease, accidents, and violence account for most mortality. For this reason few individuals in these populations ever reach the maximum possible age. Perhaps the same causes of death account for the difference between observed and pre-dicted age at death of the australopithecines.

Early hominid social ecology

With these cautions and criticisms in mind, one must accept the known fossils as the only empirical evidence we have to assess the evolution of the human patterns of growth and demography. The critical issue is not the age at death of the early hominids *per se,* rather it is their rate of growth and development. The rate of neurological development and the length of the infant and childhood dependency periods are the most important unknowns. These variables influence most strongly parental behavior toward offspring and the overall social organization and ecology of any mammal, including hominids. Dental formation is one indirect method for estimating rates of total growth and development for skeletal populations. Another method, more closely related to brain growth and neurological development, compares the size of the adult cranium to the size of the adult pelvic outlet. Living humans have the largest average pelvic outlet diameter of any hominoid. This is required in order to accommodate the birth of the relatively large-brained human infant. Even though human infants have large brains, during growth there is a tripling of brain size between birth and maturity. Estimates of pelvic outlet size exist for *Australopithecus* (Lovejoy, 1973) and early *Homo erectus* (Brown *et al.*, 1985). These fossil hominids, like living humans, have pelvic outlets about one-third the size necessary to accommodate the adult cranium. One may infer that, also like living humans, the early hominids were born at a stage of relative neurological immaturity, requiring a long period of physical dependency on more mature caregivers and an extended period for growth and development.

This does not mean that early hominids were like living humans in any other respects. It only means that certain modern human characteristics appeared early in hominid evolution. These characteristics include bipedalism, neurologically immature infants, and prolonged growth prior to sexual maturity. The early appearance of these traits is not likely to be a simple coincidence, for bipedalism, a relatively large cranial capacity in proportion to adult body size, and prolonged growth are all related to the hominid way of life. This is the way of life discussed above that includes a dependence on technology, complex and cooperative social organization, and learned patterns of behavior. Bipedalism allows for hands that are freed from locomotion and may be used to make and use sophisticated tools, gather and carry foods, and intensify social contact and communication. Slow growth and relatively large brains allow for greater learning and flexibility of behavior.

Accepting this evidence for the early evolution of the hominid pattern of physical development still leaves unresolved several issues relating to the evolution of human-like social organization. For instance, McKinley (1971) estimated that *Australopithecus* birth intervals were about equal to those for living, hunting and gathering peoples, i.e., they gave birth to a child about every three to four years. Other researchers suggest that they may have been more chimpanzee-like, giving birth about every 5.6 years. If we accept McKinley's estimates of survivorship and birth rate and Sacher's (1975) estimate of *Australopithecus* sexual maturation, then females would have their first babies at about 11 years (on average), a second infant at about age 15, a third at age 19, and would be dying, on average, about one year later. Of course, not all adults died by age 20; in fact McKinley's estimate of survivorship indicates that about 15 per cent reached 35 years of age. However, their delay in sexual maturation combined with human-like natality, but relatively early adult mortality, would result in australopithecine populations with many orphans and many more infants and children than adults. If so, one may ask, who took care of the orphaned children after their parents died? It may be argued that older *Australopithecus* adolescents, and those adults skillful, or lucky enough, to survive, must have provided the care. From this Wolpoff (1980) extrapolates that *Australopithecus* may have had social groups with human-like kinship relations, the provision of non-parental care to orphaned young, and cultural instruction of the young by both biological and social kin. This extrapolation is not unreasonable since kinship recognition and non-parental care of young by members of the social group is well known in many species of contemporary non-human primates (Jolly, 1985). Such care is only temporary in other primates; for instance, the death of a chimpanzee mother almost always leads to the death of her offspring younger than five years of age (Teleki *et al.*, 1976). Only in the human species, and possibly in other hominids beginning with *Australopithecus*, does non-parental care of the young become a long-term behavior.

Adoptions of orphaned infants by females do occur in chimpanzee social groups, but only infants older than four years and able to forage for themselves survive more than a few weeks (Goodall, 1983). Goodall noted deterioration in the health and behavior of infant chimpanzees whose mothers had died. The behavioral changes include depression, listlessness, drop in play frequency, and whimpering. Health changes were noted by loss of weight. Even those older infants who survived the death of their mother were affected by delays in physical growth and maturation. It is well known that human infants and children also show physical and behavioral pathology after the death of one or both parents (Bowlby, 1969). It seems, though, that the human infant can more easily make new attachments to other caretakers than the chimpanzee infant. The ability of the human caretaker to attach to the infant may also be greater than that of the chimpanzee adoptive caretaker. The psychological and social roots of this difference between species in attachment behavior are not known. The demographic crises in hominid evolution described above, leading to the orphaning of many young, may have supplied the selection pressures needed to further develop attachment behaviors in our ancestors.

Why non-parents, or even parents, should invest so much time and energy in any one offspring is an issue deserving its own consideration. Trivers (1972) and Dawkins (1976) analyze the evolutionary motivations for parental behavior and they conclude that the interests of parents to have a maximal or optimal number of offspring, and the interests of the offspring to extract the maximum amount of care and resources from their parents, are at conflict. The conflict in interests may be seen in terms of allocation of resources such as time and food. Women in hunting and gathering societies typically nurse their children until the age of three to four years. This places a strain on the mother, who must still find time and energy to gather food for herself and other members of her immediate family. The father, and other adults, may also be involved in providing food for the mother and her nursing infant, which limits the time these people may spend in subsistence activities, as well as in other social behavior or at rest. The mother, being physically or socially incapable of further reproduction during the nursing period, may desire to terminate the investment in her current infant as quickly as possible. If we take as given the fact that the slow growing hominid infant must take these first three to four years to acquire the physical size, abilities, and social behaviors needed to survive as a semi-independent child, there is little that mother or infant can do behaviorally to alter these conflicts of interest.

This situation is amenable to manipulation by altering growth rates. that is, one way to reduce the inevitable conflict between parent and dependent child is for the child to remain as small as possible and to grow as slowly as possible. A smaller body requires fewer nutrients than a larger body. A slow growing body needs a smaller percentage of nutrients, in relation to body size, than a fast growing body. For example, body length at birth averages about 49 to 50 centimeters for children born in the United States. The infant, at birth, may be growing in length at an average rate of up to 20 centimeters per year. Infants of average size and average rates of growth need an energy intake of about 500 calories per day, of which 23 per cent is devoted to growth. By two years of age the rate of growth decreases to about five or six centimeters per year, although now the infant is much larger than at birth, averaging abut 87 to 88 centimeters. In consequence, total caloric needs increase to about 860 calories per day, but the percentage of energy devoted to growth falls to six per cent (Payne & Waterlow, 1971). The percentage of calories needed in proportion to body length actually drops a bit as well. Slow growth, about five centimeters per year, continues until the onset of puberty. This helps to maintain childhood nutritional requirements at relatively low levels in comparison to the needs of adults, making childhood far less of an economic burden on parents than it might be if children assumed adult body size well before they achieved sexual maturity.

Case (1978) proposed another advantage of small body size in relation to slow growth. He describes people as having 'low growth efficiency,' by which he means that people have the slowest post-natal growth of virtually any mammal, but a high basal metobolic rate. For any given body size, humans require more energy for maintenance and growth than most mammals of the same body size. The relatively large brain of human children, in proportion to their body size, is one reason for the high metabolic rate. Blood circulation to the brain, supplying it with oxygen

camp boundaries while older children discharge many caretaking functions for younger children. The young children seem to transfer their attachment from parents and other adults to these older children, behaving toward them with appropriate deference and obedience . The age-graded play group functions to transmit cultural behavior from older to younger generations and to facilitate the learning of adult parental behavior (Konner, 1976). Of course, the children are never quite left on their own, as there is always one adult, or more, in camp at any time, but this person is not directly involved in child care. Rather, he or she is preparing food, tools, or otherwise primarily engaged in adult activity.

The Mbuti (nomadic hunters and gatherers of central African rain forests) have a similar child care arrangement. After weaning, toddlers enter the world of the *bopi,* the Mbuti term for a children's playground, but also a place of age-graded child care and cultural transmission. Between the ages of two or three and eight or nine, children spend almost all of their day in the *bopi.* There they learn physical skills, cultural values, and, even, sexual behavior. 'Little that children do in the *bopi* is not of full value in later adult life' (Turnbull, 1983b, p. 43–4).

The age-graded play group provides for both the caretaking and enculturation of the young, freeing the adults from these tasks so that they may provide food, shelter, and other necessities for the young, who may be at various stages of development. A women may be pregnant, have a child weaned within the past year, and have one or more older offspring simultaneously. Thus, adults may be able to increase their net reproductive output during a relatively short period of time. This benefit, and the selective advantage of a greater number of surviving offspring afforded by age-graded caretaking, may account for the evolution of the prolonged childhood and juvenile growth periods of hominids. As discussed in Chapter 2, the primate growth pattern includes a period of independence from direct maternal care (i.e., nursing) following weaning and prior to sexual maturation. This is defined as the juvenile growth stage. Contemporary humans extend this growth stage to its maximum expression of any primate. Human beings experience uniquely a childhood growth period, which spans the time between the maternal dependency of infancy and the independence of the juvenile growth stage. During the childhood growth period, human young require some care, and this may be provided by any competent adult, adolescent, or juvenile individual.

From the reconstructions of growth, maturation, natality, and mortality patterns of our hominid ancestors, prolonged juvenile growth probably first evolved with the appearance of *Australopithecus* or similar hominids. Hominid parents benefit from juveniles in terms of the child care and subsistence work they provide, which enhances the parent's prospects for further reproduction. The juveniles' potential for reproduction is also enhanced, owing to the extended period of learning prior to adulthood that slow growth and delayed maturation allow. Hominid reproduction is just not a matter of quantity of offspring, it is also a matter of quality of offspring. The ultimate goal is to produce offspring who will grow to be successful adults, that is, adults who will be able to acquire mates, and in turn reproduce and provision their own offspring. The play group, in the protective environment of the home base or camp, provides the children with the freedom for play, exploration, and experimentation, which Beck (1980) has shown encourages learning, socialization, and even tool using. Of these, learning is the most important since it is the strategy employed by hominids to ensure reproductive quality. Johnson (1982) and Lancaster (in press) show that the selective benefit of learning is that it permits adaptation to ecological changes that are not predictable. Included in these changes are common problems faced by all living and extinct hunting and gathering societies, such as seasonal variability in climate, plant growth, and animal migrations. There are also the rare crises that exemplify the unique capacity for learning in humans, such as the 1943 drought in Central Australia. Birdsell (1979) relates that during this time an old Aborigine man, Paralji, led a band of people on a 600 kilometer trek in search of water. After passing 25 dry waterholes he led them to a fallback well that the old man had not visited for more than 50 years. That well was also dry, forcing Paralji to trek 350 kilometers on ancient trails, locating water holes by place names learned from initiation rites and ceremonial songs he memorized as a child.

Even the way humans forage for food requires extensive learning during the childhood and juvenile growth periods. Much of the food humans utilize is hidden from view or is encased in protective coatings; tools are usually needed to extract and process these foods. The costs of tool manufacture, and the time and energy needed to find and process raw materials, are outweighed by the benefits. Tool-using human gatherers extract twice as many calories from savanna–woodland environments as non-tool-using primates. Other foods are poisonous before processing by washing, leaching, drying, or cooking. Acorns and horse chestnuts, eaten by many North American Indians, and manioc, a staple food of many tropical living cultures, are toxic if eaten raw. These foods must be leached by boiling in water and dried before consumption. Furthermore, knowledge of the location of these foods and their methods of processing, and the location of raw materials and their manufacture into tools, requires learning. An example is the Arunta, a hunting and gathering people of central Australia. They are compelled to live in self-sufficient nuclear families by the widely dispersed nature of food resources in their habitat (Service, 1978). There are no age-graded play groups, so children are cared for directly by adults or by their older siblings. As soon as they are able, the children follow their mothers and fathers on the daily rounds of food collection and preparation. Elkin (1964) observed boys and girls as young as five years being taken by their fathers on hunting trips and being shown how to collect raw materials and prepare them for spears, points, and other tools. Since it takes more than a decade to become proficient in the manufacture and use of these tools, early learning and slow growth and maturation are mutually beneficial.

The ecology of learning and growth

A closer analysis of the ecology of learning is required to understand how it relates to the human pattern of growth. Primates can learn by both imitation and teaching. Learning is also facilitated by social living. Jolly (1985) reviewed the data collected by primatologists conducting field studies of monkeys and apes. A common observation was that most imitative learning occurs between females and their offspring. Young baboons and chimpanzees watch their mothers select foods, peel fruits, etc. until the young begin to copy their mothers' actions. Chimpanzees learn to use tools, for instance, for termite fishing, in the same imitative way. People also learn by imitation, but purposeful teaching is equally important and, probably, more important for human beings than for non-human primates. The Arunta example of fathers teaching their children is a typical human case. People living in pre-industrial societies, and early hominids in hunting and gathering societies, survive by having available a fund of detailed knowledge of a habitat, for the location of food and raw materials and avoidance of predators. Survival also requires knowledge of intricate procedures such as detoxifying poisonous raw foods or flaking rocks into stone tools. Finally, a myriad of social facts, including knowledge of kinship organization and ritual behavior, are required. All of this information is usually passed from an older experienced individual to a younger individual. The greater the maturational status difference between teacher and pupil the more likely that the transmission of this type of detailed knowledge will be one way, older to younger. Since humans depend on complex technologies, social systems, and ideological belief systems for survival, any biological traits that facilitate the learning of cultural behavior would be naturally selected. Prolonged growth during childhood and delayed maturation fit the bill, not only because there is so much to learn, but also because maturational differences between teachers and students are maximized. This holds for the relationship between adults and children and for children in age-graded play groups.

The prolonged growth period and delay in reproductive maturity influence the development of most systems (skeletal, muscular, etc.) in the body, except the central nervous system. As was detailed in Chapter 2, the central nervous system, especially the brain, follows a growth curve that is advanced over the curve for the body as a whole. Growth in brain size is even more advanced when compared with the growth of the reproductive system; the brain achieves virtually adult size before reproductive maturation begins. Slow somatic growth and delayed sexual maturation result in a relatively small body and greater dependency of the human child during the early learning years than is the case for non-human primates. These growth patterns help establish teacher–student roles that remain stable for a decade or more, allowing a great deal of learning, practice, and modification of survival skills to occur. The advancement of central nervous system development over body growth, in general, provides the physical basis for the efficient learning and memory of these skills. It also allows for the integration of separately learned behaviors that may be creatively recombined by the individual and result in novel behaviors, suitable for situations that have never been encountered before, or changes in the physical or social environment that were not foreseen by teachers.

Learning by imitation or observation may also proceed in the opposite direction, that is, the young may introduce new behaviors to the elders of a social group. The classic field studies of Japanese macaques conducted by Itani (1958), Kawamura (1959), and Kawai (1965) clearly showed that juvenile primates are more likely than adults to investigate novel items in their environment, including new foods. It was a juvenile monkey that introduced potato washing in one group and another juvenile that introduced swimming in another group. In a similar way, human children, juveniles, and adolescents may provide a service to adults, testing new foods and behaviors for their safety and utility.

Appropriate social behavior for adult roles must also be learned, including that relating to mating and child rearing. In this regard the ontogeny of human sexual dimorphism is enlightening. There is little dimorphism in size and body composition between boys and girls prior to sexual maturation. Girls enter puberty at about the age of nine to ten years. The first outward sign of female puberty is the development of the breast bud, followed by increased fat deposition on the lower trunk (hips, buttocks, and thighs), an increase of body hair, the adolescent growth spurt in height, and finally menarche. Menarche is usually followed by a period of up to three years of adolescent sterility (anovulatory menstrual cycles), so that by the time fertility is achieved young women have completed almost all their physical growth. In contrast, boys achieve fertility well before they assume adult size and the physical characteristics of men. The genital development of boys begins, on average, only a few months after that of girls. Other physical changes of boys, the growth spurt, the deepening of the voice, and the appearance of facial hair, do not manifest themselves until genital maturity is nearly complete. Thus, the attainment of adult size in males indicates social, more than physical, reproductive maturation (Tanner, 1962; Marshall & Tanner, 1986).

Young, infertile, women may begin to advertise their reproductive potential as a means of initiating adult social interactions, including male–female bonding, sexual intercourse, and 'aunting' type child care behavior, and as a means of learning and practicing these for later in life when pregnancy and child rearing would be likely outcomes. The appearance of reproductive maturation of these older girls, and the inability of fertile adults to detect adolescent sterility, may stimulate the adults to include girls in adult social behavior. The net result is that as a primiparous mother, the mature young woman would have considerable social and parental experience, increasing the likelihood of survival of her new offspring. From an evolutionary perspective this is an important result, for the probability of death of first-born infants is higher in non-human primates than it is in humans; between 50 and 60 per cent of first-born offspring die in infancy in species of yellow baboons (Altmann, 1980), toque macaques (Dittus, 1977), and chimpanzees (Teleki *et al.*, 1976). In pre-modern human societies, such as the !Kung, about 44 per cent of children die in infancy (Howell, 1979). Just for comparison, it may be noted that in the United States, in the year 1960, about 2.5 per cent of all live, first-born children died before the age of one year (Vavra & Querec, 1973).

Boys assume adult size and appearance after reaching fertility, and therefore are perceived as adults biologically and socially at a relatively late age. This pattern of sexual development does not have the negative evolutionary impact that it would have for girls. Men contribute to the welfare and survival of their offspring, but during the period of pregnancy and lactation the male contribution is almost always mediated through the mother or other women. Men often instruct their children in technical skills (e.g., hunting) and in social or ideological behavior (e.g., religion) that may play an important role in survival. However, such instruction occurs well after the infant growth period.

There may be some positive evolutionary value in the delay young men experience between reaching reproductive maturity and later advertising this maturity in terms of physical size, the appearance of increased body hair, muscularity, and other secondary sexual characteristics. During this period of delay the endocrine changes taking place may 'prime' boys to be receptive to the roles they will have as men. Psychologists have measured changes in interests, emotions, and behavior that follow gonadal maturation, but which precede the growth spurt during adolescence. The available data are based on cross-sectional studies of European, North American and post-war Japanese youth (Schofield, 1965; Beach, 1974; Asayama, 1975). Thus, the results may be biased by sociocultural and educational factors. Nevertheless, as plasma levels of testosterone begin to increase, but before the growth spurt reaches its peak, there is an increase in psycho-sexual activity, including the onset of nocturnal emission, the intensification of masturbation, 'dating,' and infatuations (Katchadourian, 1977; Higham, 1980). There are increases in subjectively experienced feelings, such as guilt, anxiety, pleasure, and pride. Interest in adult roles and activities is heightened at this time, while simultaneously there occur alterations in psychological ties to, and behavior with, parental figures that lead toward greater independence in thinking and acting (Petersen & Taylor, 1980). Boys are becoming men, but owing to their growth spurt occurring relatively late in the maturational sequence, they may be able to learn and practice some adult behavior, including subsistence techniques, adult sociocultural roles, and sexual behavior, before they attain adult physical features and are perceived as adults by others. With such perception, young men may be expected to perform at adult levels in food production or may be considered competitors by older men in reproduction. In either case, young inexperienced men would likely not meet the expectations and challenges of other adults, leading to poor social integration and, probably, a reduced opportunity for reproduction.

Thus, girls best learn their adult social roles while they are still pre-fertile, but perceived by adults as mature. Boys may best learn their adult social roles while they are sexually mature, but not yet perceived as mature by adults. The perception of maturity is regulated by the pattern of growth in height, weight, body composition, and primary and secondary sexual characteristics. The perception of maturity is, therefore, as much subject to natural selection as are these physical characteristics.

Neoteny revisited

We are so accustomed to the positivist view of learning and teaching that most of the foregoing may appear to be obvious. Our early hominid ancestors, however, may not have seen the immediate or long-term value of parental investment, including teaching. Assuming that life for the australopithecines was relatively short, with a median age of death at 20 years, then many parents would not have lived to see the fruits of their labor, that is, the successful birth of their grandchildren. It is possible, therefore, that some biological selection took place for traits that would promote the provisioning of children and their instruction in survival skills and other forms of parental investment, without conscious knowledge of the reasons for this behavior. These traits might, ideally, be closely linked with the pattern of growth during infancy and childhood: the time when parental investment in children, their provisioning and instruction, plays a crucial role in immediate and long-term survival.

The concept of neoteny, criticized in the last chapter as the mechanism for human growth, may have some relevance in terms of parental investment. Lorenz (1971) contends that the physical characteristics of mammalian infants, including small body size, a relatively large head with little mandibular or nasal prognathism, relatively large round eyes in proportion to skull size, short thick extremities, and clumsy movements, inhibit aggressive behavior by adults and encourage their caretaking and nurturing behaviors. Lorenz believes that these neotenous features trigger 'innate releasing mechanisms' in adult mammals, including humans, for the protection and care of dependent young. As evidence for the genetic cause of human parental behavior toward neotenous children, Lorenz cites the fact that people extend nurturing behavior towards other animals with appropriate features, e.g., puppies, kittens, and even adult animals with neotenous features (Figure 3.1). Lorenz notes the penchant of some pet owners to treat the miniature varieties of dog breeds as if they were infants, even when the animals are mature. Miniaturization in domesticated animals results from a growth process that retains neotenous features (e.g., large round head and short extremities relative to body size) into adulthood. Lorenz further notes that inanimate objects such as clouds and rock formations that resemble neotenous animals can release emotions of tenderness from humans.

Gould (1979) questions the innateness of the human response to infantile features. Such behavior may be 'learned from our immediate experience with babies and grafted upon an evolutionary predisposition for attaching ties of affection to certain learned signals' (p. 34). Whether innate or learned the resultant behavior is the same. The depth of our reaction to neotenous features is exemplified by Gould's description of the 'evolution of Mickey Mouse.' From the cartoon creature's first appearance in 1928 to later appearances in the 1950s, Mickey's body shape changes. His head enlarges, eyes enlarge, and arms and legs shorten relative to body size. During this physical evolution Mickey is transformed socially from an ill-behaved, mischievous rodent, to the delightful host of the Magic Kingdom. The physical and behavioral

changes are not simply coincidental, rather they are necessary concomitants of human perception based upon our reaction to body shape and
growth status.

An elegant series of psychological experiments performed by Todd *et
al.* (1980) confirm that human perceptions of body shape and growth
status are consistent between individuals. When about 40 adults (from the
University of Minnesota community) were shown a series of profiles of
human skull proportions, they could easily arrange them correctly into a
hierarchy spanning infancy to adulthood. The adults could also ascribe
maturity ratings to skull profiles that were geometrically transformed to
imitate the actual changes that occur during growth. This perception was
selective because a variety of other types of geometrical transformations
elicited no reports of growth or maturation (Figure 3.2). When the
growth-like mathematical transformations were applied to profile drawings of the heads of birds and dogs, human subjects reported identical

Figure 3.1. The releasing schema for human parental care responses. Left: head
proportions perceived as 'loveable' (child, jerboa, Pekinese dog, robin). Right:
related heads which do not elicit the parental drive (man, hare, hound, golden
oriole) (from Lorenz, 1971).

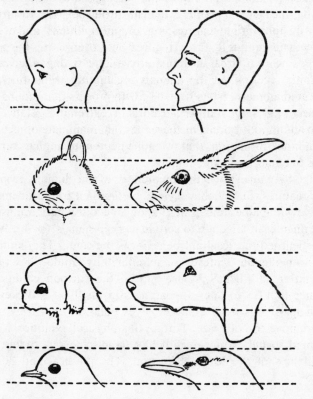

perceptions of growth and maturation, even though in reality the development of these animals does not follow the human pattern of skull shape change. Even more surprising is that subjects reported the perception of growth when the growth-like mathematical transformations were applied to front and side view profiles of Volkswagen 'beetles,' objects which do not grow.

In another series of experiments, Alley (1983) studied the association between human body shape and size, and the tendency by adults to protect and 'cuddle' other individuals. The subjects of the experiments were 120 undergraduate students, 45 men and 75 women, at an American university, ranging in age from 17 to 27 years (mean age = 18.8 years). All the subjects were childless, and about one-third had no younger siblings. In the first experiment, subjects were shown two sets of drawings. One set was based upon the diagrams of human body proportion by Stratz (1909) presented in Figure 1.3. Alley's version of these diagrams, which are reproduced in Figure 3.3, were called 'shape-variant' drawings. He used the middle-most of the profiles in Figure 3.3 (the 'six-year old' profile) to construct a series of figures that varied in height and width, but not in shape. Figure 3.4 is an example of one of Alley's size-variant drawings. Note that, unlike the Stratz diagrams, Alley's figures have no facial features, navels, or genitals. Perceptual differences between figures would be due to body shape or size alone, not sex, nasal prognathism, relatively large round eyes, or the stage of sexual maturation.

Figure 3.2. Two of the mathematical transformations of human head shape used in the experiments of Todd *et al.* (1980). The middle profile in each row was drawn from the photograph of a ten year old boy. The transformations were applied to this profile of a real child. The cardioidal strain transformation is perceived by most adults as growth. The affine shear transformation is not perceived as growth.

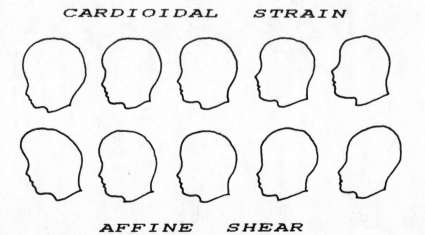

CARDIOIDAL STRAIN

AFFINE SHEAR

The subjects were shown pairs of the shape-variant drawings (i.e., profiles of a newborn and a six-year old, a two-year old and a 12-year old, etc.) or pairs of the size-variant drawings and asked to state which one of the pair they 'would feel most compelled to defend should you see them being beaten.' In 10 trials, the mean reponse was 7.5 (sd = 2.7) for the 'younger' of the shape-variant drawings and 8.7 (sd = 2.1) for the smaller of the size-variant drawings. In the second and third experiments, the same subjects were shown the five shape-variant drawings, one at a time, and asked to rate the drawings on a scale of from 1 (low) to 9 (high) 'according to how compelled you would feel to intervene if you saw someone striking the human depicted' and their feelings to 'hug or cuddle' the person depicted. The results of both experiments are given in Table 3.2. The trends found in both experiments are statistically significant. There was a fairly strong reported willingness to defend 'newborns' and 'two-year olds,' and a moderate willingness to defend 'older' persons. The reported willingness to cuddle decreased with the 'age' of the drawings. Also, the desire to cuddle was not as strong, and, statistically significantly less well ordered, than the willingness to defend. Alley mentions that American cultural values (e.g., defend the 'underdog' or the weak) may have influenced the results of these experiments. However, placed in the context of the ethological study of 'parental' caregiving in mammals and birds, Alley believes that his results demonstrate a more

Figure 3.3. The series of five shape-variant drawings used in the experiments of Alley (1983). These drawings show the typical body proportions of a male at (from left) birth, two, six, 12, and 25 years of age.

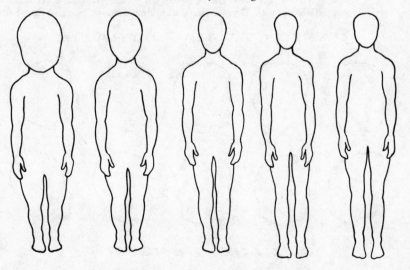

Table 3.2. *Mean reported willingness to defend or cuddle persons of different body proportions (Alley, 1983)*

Age portrayed (years)	Defend	Cuddle
Newborn	7.7 (1.7)	3.4 (2.3)
2	7.1 (1.7)	3.4 (2.0)
6	5.8 (1.8)	3.2 (2.2)
12	5.0 (1.9)	3.0 (1.9)
25	4.3 (1.9)	2.7 (2.2)

Note: Standard deviations are given in parentheses.

Figure 3.4. An example of the size-variant pairs of drawings used in the experiments of Alley (1983).

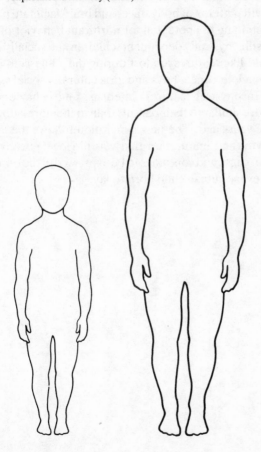

general tendency to protect or cuddle others based on the perception of maturational status.

The psychological experiments of Todd *et al.* and Alley provide support for the arguments developed in this chapter for the evolution of the human pattern of growth and maturation. Perhaps other experiments could be designed to test the hypotheses relating to the evolution of different patterns of adolescent maturation followed by boys and girls.

The human infant's morphological profile is neotenous. Human children, and probably the children of ancestral hominid species, retain neotenous characteristics longer than any other mammalian species. This is due, in part, to the slow growth of most parts of the body, especially post-cranial regions and also parts of the head such as the nose and the jaw (due largely to delayed molar tooth eruption). The perception of neoteny is also aided by the relatively faster growth of the brain compared with the face and body, which maintains the round, bulbous cranium of infant mammals throughout childhood and into the juvenile period of human growth. These growth patterns of body, face, and brain facilitate parental investment by maintaining the potential for nurturant behavior of adults towards older, but still physically dependent, children and socially dependent juveniles. Child-like features are lost during the adolescent growth spurt. At the end of adolescence, boys and girls enter the social world of men and women. In physical features, interests, and behaviors these young adults are more similar to their parents than to their pre-adolescent selves of just a few years ago. The new generation follows the cycle of reproduction, growth, and maturation that was phylogenetically set in place millions of years ago, and continues to be expressed in the ontogenetic development of every human child born today.

4 Growth variation in living human populations

Population differences in stature, body weight, and other physical dimensions have been documented throughout recorded history. Ancient Egyptian sources mention groups of very short stature people living near the headwaters of the Nile River, possibly ancestors of central African 'pygmy' populations alive today (Hiernaux, 1974). Museum displays of medieval armor and fashion often provoke visitors to comment on how much bigger European people are today than in the past. Human biologists have recorded the variation in size that exists between living populations, and found that it is relatively easy to describe the differences in size, but much more difficult to explain why this variation exists. Population differences in body size, including variation in amounts and rates of growth, are due to a wide range of hereditary and environmental factors. Examples of some of these were given in Chapter 1. The present chapter begins with a review of some studies of population variation in growth and development and follows with a discussion of the interaction between hereditary and environmental causes of such variation. A consideration of the evolutionary value of population variation in body size completes this chapter.

Population differences in growth and development

The average height, weight, and weight/height (a simple measure of body proportion) for several human populations are given in Table 4.1. The data are listed in descending order according to the average height of the men, although with the exception of only the Aymara, the heights of the women follow the same descending order. Young adults in the Netherlands may be, on average, the tallest people in the world. Some of the shortest young Dutch men, those at the third percentile, have a height of 169.3 cm and, thus, are about equal to the mean height of men in Japan, and taller than the mean stature of men in Guatemala and Bolivia. Young adults in the United States are, on average, shorter than the Dutch, but, relative to the other populations, are a 'tall' group of people. The Turkana, who are nomadic, animal-herding pastoralists living in rural Kenya, rank next in average height. The sample of Japanese represent reasonably affluent university students. They are the tallest and heaviest,

Table 4.1. *Average height (cm), weight (kg), and weight/height of young adult men and women in several populations*

Population	Age (years)	Height Men	Height Women	Weight Men	Weight Women	Weight/height Men	Weight/height Women
Netherlands, national sample (medians)	20	182.0	168.3	70.8	58.6	0.39	0.35
United States, national sample (medians)	20–21	177.4	163.2	71.9	57.2	0.41	0.35
Africa, Turkana pastoralists (means)	20	174.3	161.6	49.8	47.4	0.29	0.29
Japan, Tokyo students (means)	20	169.7	157.3	64.0	51.8	0.38	0.33
Guatemala, Maya Indians (means)	17–18	167.0	143.5	52.4	45.7	0.31	0.32
Bolivia, Aymara Indians (means)	20–29	162.0	149.0	58.1	52.4	0.36	0.35
Africa, Mbuti Pygmy (means)	Adult	145.0	138.0	40.0	37.0	0.28	0.27

Sources: Netherlands, Roede (1985); United States, Hamill *et al.* (1977); Africa, Turkana, Little *et al.* (1983); Japan, Kimura (1984); Guatemala, Mendez & Behrhorst (1963) and Sabharwal *et al.* (1966); Bolivia, Mueller *et al.* (1980); Africa, Mbuti, Mann *et al.* (1962).

on average, of any group of young Japanese adults measured this century, but they are considerably shorter and lighter, on average, than the Dutch, the Americans, or the Turkana.

The Maya of Guatemala and the Aymara of Bolivia are native American peoples. Both groups live in rural areas, practice subsistence farming and herding, and many of them suffer from mild-to-moderate malnutrition. Low economic status and undernutrition are associated with growth retardation, and these are likely to be factors that account for the relative short stature of the Maya and Aymara.

Finally, African Mbuti pygmies may be, on average, the shortest people in the world, and their short stature appears to have a strong genetic component (Rimoin *et al.,* 1968; Merimee *et al.,* 1981). However, there is a wide range of variation in the stature of individual pygmies. Barnicot (1977) compared the distribution of male stature of the pygmies with another African people, the Tutsi, who have a mean stature of 176.5 cm, and found that the tallest pygmy men were larger than the shortest Tutsi men. This analysis shows that average figures may be quite misleading for individuals within a population. Even so, the statistics for average size given in Table 4.1 indicate patterns of growth that are useful for descriptive purposes, and provide a starting point for the analysis of the causes of such variation.

The rank order of mean weights in Table 4.1 does not follow the same order as stature. On average, United states men are the heaviest and Mbuti women are the lightest of all the populations listed. The relatively tall Turkana have an average weight less than that of any other sample save the Mbuti, and the Bolivian Aymara are heavier than the taller Guatemalan Mayans.

The ratios of weight for height show that the Turkana and the Mbuti have the lowest values, reflecting their linear body build. The similarity in the proportion of height to weight between these African samples is striking, since the Turkana are on average more than 25 cm taller than the Mbuti pygmy sample. Dutch and American men have the highest average ratios, meaning that there is, on average, relatively greater weight for height in these two populations than in the other samples of men. Little *et al.* (1983) compared the Turkana with a United States reference population and found the greater weight for height of the Americans was due to both more fat and more lean tissue (e.g., muscle), but especially more fat.

Rates of growth also vary considerably between human populations. In Figure 4.1, the velocity of growth in height is presented for three

Figure 4.1. Mean constant curves, estimated by the Preece-Baines model 1 function, for the velocity of growth in height of British boys, Australian Aborigine boys, and African boys. The curves were drawn using the 'Growth Package Programs' and data contained therein (Brown, 1983).

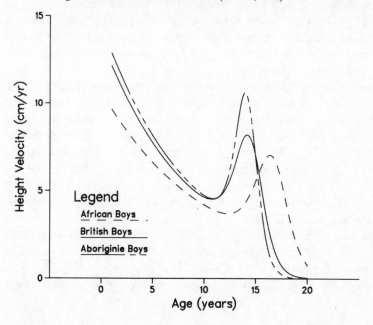

populations: a sample of British boys (Preece & Baines, 1978), a sample of boys from The Gambia, West Africa (Billewicz & McGregor, 1982), and a sample of Australian Aborigine boys (Brown & Townsend, 1982). The velocity curves were calculated from longitudinal measurements of height by fitting the measurements to the Preece–Baines model 1 function using algorithims developed by Brown (1983). This mathematical function estimates the mean constant velocity of growth from childhood through the attainment of adult height. The function also estimates several biological parameters of growth that occur during the adolescent growth spurt, for instance, age at the beginning of the spurt, age at the peak of the spurt, and the velocities of growth at these ages. Assessments of the Preece–Baines model 1 function find its statistical applications to be accurate and reliable (Marubini & Milani, 1986).

Compared with the British boys, the Australian Aborigine boys have a greater velocity of growth at the peak of the spurt (10.6 cm/yr compared with 8.2 cm/yr), but differ only slightly in other aspects of the shape of the velocity curve and the timing of adolescent growth events. The Australian boys are shorter than the British boys at all ages prior to the onset of adolescence, and despite their greater peak velocity during the spurt the Australians remain shorter than the British at adulthood (174.6 cm compared with 172.1 cm; Brown & Townsend, 1982). The onset of the adolescent spurt for the African boys occurs about 1.5 years later than that for the British or Australian boys. The Africans also have a more prolonged spurt, taking 4.08 years to reach the peak of the spurt compared with an average of 3.29 years for the British and Australian boys. The African boys have the lowest value for growth velocity at the peak of the spurt (6.9 cm/yr), and despite their relatively longer period for growth their final height, at 170.77 cm, is less than that for the British or Australian boys (Brown, 1983).

Other aspects of biological maturation may differ from one population to another. Masse & Hunt (1963) found that the dental and skeletal developments, as measured by radiographs of tooth formation and of the appearance of ossification centers in the hand and wrist, of African blacks were, on average, more advanced at birth than that of European whites. However, by two to three years of age the Africans fell behind the Europeans in these developmental measures. Jones & Dean (1956) and Garn & Bailey (1978) suggested that African blacks are 'genetically programmed' to develop more rapidly than European whites, however a postnatal delay in development occurs in the Africans, compared with the Europeans, due to an adverse nutritional, disease, and socioeconomic environment for growth of the African children.

When children of European ancestry (whites) and predominantly

African ancestry (blacks) living in the United States are matched for socioeconomic status variables, such as mother's occupation and education, it is found that black children are consistently advanced over white children in the formation and emergence of the permanent teeth, and advanced in radiological appearance of ossification centers of the skeleton and epiphysial union (Garn & Bailey, 1978). The precocity in dental and skeletal development of black over white children can also be observed during the prenatal period.

Hereditary and environmental causes of population variation

The developmental advancement of American black children compared with American white children during prenatal and postnatal life, especially when the families of the children are matched for socioeconomic status, indicates a strong genetic control over some aspects of the maturational process. However, starvation, disease, and, generally, poor living conditions of many black children in Africa and elsewhere in the world exert a substantial influence that may delay growth and development. In reality, the biological development of the human being is always due to the interaction of both genes and the environment. It is erroneous to consider whether one or the other is more important; genes are inherited and 'everything else is developed' (Tanner, 1978, p. 117).

Comparisons of stature and body proportion between blacks and whites in the United States provide an example, and convenient starting point, for the discussion of gene–environment interactions and their effect on growth. Fulwood *et al.* (1981) published data from the first National Health and Nutrition Examination Survey (NHANES I) of the United States, which gathered anthropometric data on a nationally representative sample of blacks and whites aged 18 to 74 years. When the data are adjusted for differences between the two ethnic groups in income and education, urban or rural residence, and age, there is no significant difference in average height between black and white men. Nor is there a significant difference in average height between black and white women.

Although white and black adults in the United States have the same average stature, when education, income, and other variables are controlled, the body proportions of the two groups are different. Krogman (1970) found that for the same height, blacks living in one American city had shorter trunks and longer extremities than whites, especially the lower leg and forearm. Hamill *et al.* (1973) found that this was also true for a national sample of black and white youths 12 to 17 years old. A genetic cause for the body proportion differences between blacks and whites seems likely, since the samples were matched for major environmental determinants of growth and statures did not differ.

Differences in body proportion are known from other populations. Eveleth & Tanner (1976) surveyed studies of boys and girls of European (London), African (Ibadan), Asian (Hong Kong), and Australian Aborigine origin. In proportion to sitting height, the Australians had the longest legs followed, in order, by the Africans, Europeans, and Asians. Expressed quantitatively, 'at a sitting height of 60 cm, for example, London boys have leg lengths averaging 43 cm, Ibadan boys 53 cm and Australian Aborigine boys 61 cm' (Eveleth & Tanner, 1976, p. 229). Generally, it is assumed that the body proportion differences between geographic populations are explainable only in terms of a genetic model, though the mechanism is not known.

Recent work on the growth of Japanese children suggests that environmental factors also may be powerful determinants of body proportion. Kondo & Eto (1975) found that between the years 1950 and 1970 the ratio of sitting height to leg length decreased for Japanese school children, meaning that the children became relatively longer legged with time. Tanner *et al.* (1982) confirmed this finding by comparing both the rate of growth and the amount of growth for Japanese school children measured in 1957, 1967, and 1977. Each successive cohort of children grew faster and larger than the previous cohort. Between 1957 and 1977, sitting height showed practically no increase, while increased leg length accounted for almost all of the difference in height (4.3 cm for boys and 2.7 cm for girls). In 1977, adult Japanese had sitting height to leg length proportions similar to northern Europeans, whereas 20 years before the two populations were significantly different. The major influences on growth that changed during the past two decades are those related to the environment, including nutrition, health care, sanitation, and socioeconomic status (Kimura, 1984; Takahashi, 1984). In average stature, the Japanese remained shorter than Europeans by about 5.5 cm. Kimura (1984) states that the trend for increased stature in Japan has stopped. If so, the major genetic difference in growth between contemporary Japanese and European populations is more likely to be for total size, rather than for body proportions.

Another approach to the study of population differences in growth and development is to compare children and adults of different national or geographic backgrounds living in the same, or very similar, environments. The differences found may indicate an hereditary determination of amount and/or rate of growth. Ashcroft & Lovell (1964) and Ashcroft *et al.* (1966) measured the heights and weights of four- to 17-year old children and youth of European, African, Afro-European, and Chinese backgrounds living in Kingston, Jamaica. All the children were from upper-middle to upper socioeconomic status and attended private fee-paying schools. There were no significant differences in height or weight between the European, African, and Afro-European groups. However, the Chinese sample were significantly shorter and lighter than the other three groups at almost every age, suggesting an hereditary difference in amount or rate of growth between the Chinese children and the other samples of children.

Mexican–American and white (European ancestry) adults living in three northern California communities were studied by Stern *et al.* (1975). The residents of each community were matched for socioeconomic status variables. Within each community the Mexican–Americans were significantly shorter than the whites. Similar results were found for samples of adult Mexican–Americans and whites living in Texas (Malina *et al.*, 1983). The Texas study compared samples at two different levels of socioeconomic status, and the Mexican–Americans were shorter than the white Americans, on average, within both levels of socioeconomic status. The differences in stature within each level may still have been caused by environmental factors that affected childhood growth. However, a statistical analysis showed that the difference in height associated with Mexican or European ancestry, holding socio-economic status constant, averaged 5.6 cm and the difference due to lower or higher socioeconomic status, holding the effect of ancestry constant, was only 1.4 cm. It seems reasonable to propose that genetic factors contributed to the difference in height between the Mexican–Americans and white Americans.

Social and biological factors associated with the height of children of different national and cultural origin living in England were studied by Rona & Chinn (1986). A sample of 13 107 boys and girls aged five to 11 years, living in low socioeconomic status communities, especially inner-city areas, were measured in 1983. The children were of Afro-Caribbean, Indian–Pakistani, and European background. Mean heights of children from these three groups were compared with a nationally representative sample of British children measured in 1982. It was found that the Afro-Caribbean children were on average 3.5 cm taller than the national sample, while the Indian–Pakistani and European children were on average about 2.5 cm shorter than the national sample. Statistical adjustments of the mean height values for the effects of parental social class, employment, time of migration to England, nutritional factors, number of children in the family, and other social and biological variables did not change the pattern of results. The population differences in height for lower socioeconomic status children in England parallel values for the average heights of Africans, Europeans, and Asians living on their native continents (Eveleth & Tanner, 1976). This suggests a major hereditary contribution to the pattern of stature differences between geographic populations, although it is clear that environmental conditions for growth contribute to the actual mean stature achieved by any population.

These studies, however, cannot be summarized in a simple manner, and it is very difficult to sort out the relative contributions of genes and the environment to growth. Controlled experimental studies of human development are not possible to conduct, legally and morally. Other types of naturalistic observations and more detailed analysis appear to be needed to sort out the factors that determine population variation in growth.

Bogin & MacVean (1978; 1982) studied the growth of Guatemalan school children between the ages of seven and 14 years. Children of Guatemalan Ladino ethnicity and European ancestry participated in the studies. Ladinos were defined as those children with both parents and all four grandparents born in Guatemala and possessing Spanish surnames. Europeans were defined as those children with both parents and all four grandparents born in Europe (excluding Spain and Portugal) or North America and possessing European (non-Spanish) surnames.

One group of Ladino and European children attended a private school, which, because of its fee structure and admission requirements, was virtually restricted to children of very high socioeconomic status. The nutritional status, health care, physical and psychological stimulation, social and academic education, and other correlates of physical growth of these children are known to be equal to those of well-off populations in developed nations (Johnston *et al.*, 1973; Bogin & MacVean, 1983). Growth data for these high SES children are presented in Table 4.2, and indicate that, at seven years of age, European boys and girls were significantly taller than Guatemalan boys or girls. There were no significant differences in weight or skeletal 'age' (an indicator of biological maturity) between the two groups. Weight is usually considered to be more sensitive to environmental influences than height (Waterlow *et al.*, 1977), and the equality in weight between the two groups may reflect their common, favorable environment for growth. Since the children were maturing at the same average rate, the significant difference in height between the European and Guatemalan children, despite the common environment, indicates a possible genetic difference in linear growth between the two groups.

Further support for a genetic interpretation of the difference in height between European and Guatemalan children comes from a study conducted by Johnston *et al.* (1976). This study used longitudinal data, from children aged five to 17 years old, to compare the pattern of growth of Ladinos with that of Europeans. Both samples of children lived in

Table 4.2. *Heights, weights and skeletal ages*[1] *of seven-year old European and Ladino boys and girls of high socioeconomic status living in Guatemala*

	Europeans		Ladinos	
	Boys	Girls	Boys	Girls
Sample sizes	171	158	412	276
Height (cm)				
mean	125.59	123.32	124.16	121.71
sd	4.80	4.79	4.80	4.80
Weight (kg)				
mean	25.75	25.20	25.36	24.83
sd	4.74	5.07	3.95	4.25
Skeletal age (yr)				
mean	6.84	7.22	6.93	7.30
sd	1.19	0.91	1.09	0.84

Note: Significant differences exist between European and Ladino boys and girls in height; no differences, by sex, for the other growth variables (from Bogin & MacVean, 1982).
[1]Skeletal ages were estimated from hand–wrist radiographs using the method of Greulich & Pyle (1959).

Guatemala and were of upper socioeconomic status. Johnston *et al.* used a mathematical model of growth (Bock *et al.*, 1973) to analyze the data. The model takes the longitudinal growth data for a child, and separates the total set of data into a pre-adolescent and an adolescent component. In this way, the adult difference in height between individuals, or samples of people, may be ascribed to different amounts and rates of growth during childhood, during the adolescent growth spurt, or during both growth periods. Johnston *et al.* also included a sample of children of European descent, and of upper-middle socioeconomic status, living in California. The California sample was included so that the growth of a group of European children, living in a non-Guatemalan environment, could be compared with the growth of the Europeans living in Guatemala.

The results of the study are presented in Table 4.3. Prior to adolescence, the European and Guatemalan children living in Guatemala had no significant difference in amount of growth, but both groups grew significantly less than the Europeans in California. During adolescence, the amount of height achieved by both groups of European children was significantly greater than the height added by the Guatemalan Ladino children. There was no significant difference between the European groups in the average amount of growth during adolescence. The result of this pattern of growth was that the mean adult height of Europeans living in Guatemala was intermediate between that of the shorter Guatemalan

Table 4.3. *Estimated mean values and standard deviations (sd) for the pre-adolescent and adolescent components of longitudinal growth in height (cm) of Guatemalan and European children living in Guatemala and European children living in the United States*

	Pre-adolescent		Adolescent	
	mean	sd	mean	sd
Guatemalans in Guatemala				
Males (20)[1]	147.1	6.0	26.8	3.0
Females (15)	141.2	6.3	20.7	6.0
Europeans in Guatemala				
Males (14)	145.4	9.0	29.9	6.7
Females (9)	138.7	7.0	25.7	4.2
Europeans in California				
Males (65)	151.4	6.7	29.5	5.3
Females (64)	143.0	6.7	25.1	4.4

[1]Sample size given in parentheses.
Source: Johnston *et al.* (1976).

Ladinos and the taller Europeans in California. From these results, the authors of this study hypothesized that, prior to adolescence, growth is relatively more sensitive to the environment; hence, children of different populations, with potentially different genetic backgrounds, living under similar environmental conditions will tend to be of the same average size. During adolescence, however, genetic determinants of growth are more strongly expressed, and population differences in growth will be seen despite a common environment. Adolescent youths of similar genetic backgrounds, even if living in different environments, will tend towards similar amounts of growth.

Other studies have noted the sensitivity of the childhood growth period to environmental modification and the resistance of the adolescent phase of growth to such influences. Garn & Rohmann (1966) and Hunt (1966) analyzed longitudinal data from growth studies conducted in the United States and found that parent–child correlations in height were lower during the pre-adolescent years than during adolescence. Higher correlation in parent–child growth during the child's adolescence may indicate a greater expression of genetically inherited growth potential, but it may also be a reflection of the greater length of time adolescents have spent living with their parents, compared with pre-adolescent children. Thus, an environmental effect cannot be ruled out in this type of analysis.

Frisancho *et al.* (1980) compared the growth of Quechua Indian and

Mestizo children, aged six to 19 years old, living in the same town in a lowland province of eastern Peru. A genetic difference between the Indians and the Mestizos was suggested by finding statistically significant differences in their mean values for ABO and Rh blood group antigens and skin reflectance measurements. The children of both groups were matched for nutritional status, and it was found that both groups attained the same mean height during childhood, but that during adolescence the Mestizos were taller, on average, than the Quechua. The authors also identified adolescents with evidence of chronic undernutrition, and found that their height was significantly below that of better nourished adolescents, be they Mestizo or Quechua. It was concluded that the common environment seemed to affect equally the childhood growth of both ethnic groups, but during adolescence, genetic differences in amounts of growth were expressed. However, in some of the children an environmental factor, chronic undernutrition, interacted with the genes for growth of both samples of children, and reduced the magnitude of the typical ethnic differences in the pattern of growth expressed during puberty.

Kimura (1984) reviewed research that was conducted in Japan on the offspring of Japanese–American parents (typically, Japanese women and American white and black servicemen). During childhood (six to 11 years of age), the Japanese–American children grew, on average, about the same amount as native Japanese children in height, weight, and most other body measurements. By age 15, the Japanese–American children were larger than the native Japanese in height and weight, but were shorter and lighter than black and white children living in the United States. Thus, on both sides of the Pacific Ocean the environment seems to be a relatively strong determinant of amounts of growth during the childhood years, while heredity has a relatively strong influence on adolescent growth.

Given the consistency in the results of these studies, based on populations from Guatemala, the United States, Japan, and Peru and from diverse cultural and socioeconomic backgrounds, it seems reasonable to accept their common findings. However, there are exceptions to the general rule that the environment is the primary determinant of pre-adolescent growth, for example, the studies by Ashcroft & Lovell (1964) and Ashcroft *et al.* (1966) in Jamaica, and Bogin & MacVean (1982) in Guatemala, described above. In these three studies, some samples of pre-adolescent children of different geographic and cultural origin, but living in a similar environment, were found to differ significantly in stature.

Population differences in body composition have been found in several studies. Piscopo (1962), Malina (1966), Johnston *et al.* (1974a; 1974b) and

Harsha *et al.* (1980) found that, in the United States, black children and youths have, on average, less total subcutaneous fat than white children and youths, and that the blacks have relatively less subcutaneous fat on their extremities compared with the whites. Robson *et al.* (1971) measured the triceps and subscapular skinfolds of children of African (black) descent, one month to 11 years old, living on the island of Dominica. Only healthy, well-nourished children were included. The data were compared with similar measurements taken from a sample of English children of the same ages (the ethnic composition of the English sample was not specified, but was implied to be of mostly European background). The black childen of Dominica were, on average, leaner than the English children, and this difference was entirely due to the Dominicans having significantly smaller triceps skinfolds. There were no significant differences between the populations in the mean subscapular skinfold thickness. The results of these studies suggest that during childhood and adolescence, blacks have less total body fat, and a different anatomical distribution of subcutaneous fat, than whites.

Similar average differences in fatness and fat distribution have been found when other samples of African black children and American or European white children were compared (Eveleth & Tanner, 1976). It is possible to conclude from these studies that an ethnic (i.e., genetic) difference in mean fatness and typical fat distribution may exist between children of African and European origin. Of course, many individual children do not follow the mean tendency for body composition of their natal population. Moreover, Mueller (1982) showed that racial or ethnic differences explain relatively little of the mean differences in body composition between blacks and whites living in the United States. In this study, Mueller used multivariate statistical analysis to identify mathematically factors that influence fatness and fat distribution. A factor Mueller called 'ethnicity' was found to be statistically significant, but accounted for only about two per cent of the variance in fatness and five per cent of the variance in fat distribution between the two samples. Sex, age, and unspecified factors, of both genetic and environmental origin, accounted for most of the variance in fatness and fat distribution.

Johnston *et al.* (1975) found no evidence for an ethnic or genetic determination of fatness or fat distribution for samples of children of European ancestry and Guatemalan Ladino ancestry (as previously defined) living in Guatemala. Both groups of children were of high socioeconomic class, attending the same private school, so in several ways they were exposed to a common environment. As measured by skinfolds, there were no significant differences in fatness between the ethnic groups at either the triceps or subscapular skinfold sites. Since the triceps site correlates highly with other measures of extremity fatness and the subscapular site correlates highly with other measures of trunk fatness, it may be inferred that there is little ethnic difference in fat distribution in this case. Johnston *et al.* suggested that the common pattern of fatness and fat distribution of the Europeans and Guatemalan Ladinos might be due to their living under similar environmental conditions.

Bogin & Sullivan (1986) found that the environment is the major determinant of the average amount of fatness and the fat distribution of children. The authors compared the fatness and fat distribution of four groups of children, aged seven to 13 years old, living in Guatemala. The groups (and sample sizes) were: Guatemalan Ladinos of high socioeconomic status (SES) (320 children), Europeans of high SES (164 children), Guatemalan Ladinos of low SES (340 children), and Guatemalan Indians of very low SES (669 children). Triceps and subscapular skinfolds were measured for each child. Previous research with these same groups had shown that SES was significantly associated with nutritional status, as reflected by skinfolds and other measures of body composition (Bogin & MacVean, 1981a; 1981b; 1984). So, as expected, and as shown in Figure 4.2, the high SES Ladinos and Europeans had, as a group, larger skinfolds than the low SES Ladinos or very low SES Indians.

It was hypothesized that socioeconomic status would also be associated with fat distribution. Trunk fat may be more physiologically important than extremity fat, for example, trunk fat may serve to protect the internal organs, and is associated with reproductive development in women. Accordingly, if total body fatness is reduced in children from lower socioeconomic status populations, there should be a relatively greater reduction of extremity fat and a relatively greater retention of trunk fat. As shown in Figure 4.3, Bogin & Sullivan found that high SES Ladinos and high SES Europeans in Guatemala had a similar distribution of fat between the triceps and subscapular skinfold sites. The low SES Ladinos had significantly less fat at the triceps site than the two high SES groups, and the very low SES Indians had significantly less fat at the triceps site than any of the three other groups. Thus as SES decreased, fat distribution became more centripetal, that is, relative amounts of arm fat decreased and relative amounts of trunk fat increased. In accordance with the hypothesis, the relative fat distribution changed even as the absolute amount of subcutaneous fat decreased at both arm and trunk sites from high, to low, to very low SES. Thus, fat distribution in these samples of children was determined, at least partly, by socioeconomic status, which is a proxy measure for nutritional adequacy and other environmental variables. No evidence for a genetic or an ethnic effect could be demonstrated for these samples of children.

Adult differences in body composition have been studied in several populations. Johnson *et al.* (1981), using data from a national sample of the United States, found that the average black–white differences in fatness and fat distribution persisted into adulthood for males, from one to 74 years of age. That is, at all ages and at all levels of fatness, black males were leaner, and especially so on the extremities, than white males. Black females were found to be leaner, on average, than white females from the ages of one to 24 years, but from age 25 to 74 years, black females had larger mean values for the triceps and subscapular skinfolds than white females. This suggests that as adults, black women are, on average, fatter than white women, and that the relative anatomical distribution of fat of the black women has changed since childhood.

Fat distribution was studied in Mexican–American and white American adults living in California (Stern *et al.*, 1975) and Texas (Malina *et al.*, 1983). The subjects were matched for socioeconomic status variables. The two samples did not differ in the mean size of the triceps skinfold, but Mexican–Americans had, on average, significantly larger subscapular skinfolds than the whites.

A study by Malina *et al.* (1982b) compared athletes participating in the

Figure 4.2. Mean values for the sum of the triceps and subscapular skinfolds of children living in Guatemala. The raw data were log transformed to normalize the distribution of the values. Larger log values indicate greater fatness. The high SES samples of children are significantly fatter than the low SES samples of children (from Bogin & Sullivan, 1986).

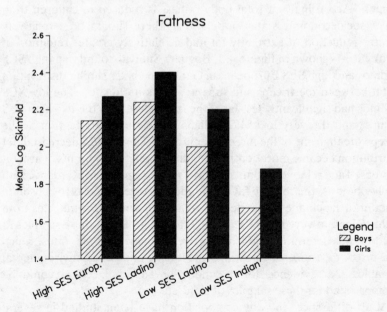

Montreal Olympic games of 1976 for skinfold measures of subcutaneous fatness and fat distribution. The sample included 264 white male and 133 white female athletes, 38 black male and 10 black female athletes, and seven Asian male and four Asian female athletes, representing 46 countries, 20 major sports, and 68 different Olympic events. The median age of the male athletes was 21 years and that of the female athletes was 23 years. It was found that the black athletes were significantly less fat than the white or Asian athletes. The total variance in fatness was statistically partitioned into the following factors (and percentage of variance): sex (31 per cent), sport (19 per cent), ethnicity (3 per cent), and age (3 per cent). The other 44 per cent of the variance in fatness is not explained by these factors and must be due to other (uninvestigated) causes. In terms of fat distribution, white athletes had significantly more fat located on their extremities than Asian athletes, who had more fat located on the trunk of their bodies. The fat distribution of the black athletes was intermediate between that of the white and Asian athletes.

Figure 4.3. Relative distribution of subcutaneous fat at the triceps and subscapular skinfold sites for children living in Guatemala. Fat distribution is expressed as principal component scores; larger scores indicate relatively greater triceps fatness, smaller scores indicate relatively greater subscapular fatness. The low SES samples of children have significantly more subscapular fat, relative to triceps fat, than the high SES samples. No significant difference exists between the two high SES samples (from Bogin & Sullivan, 1986).

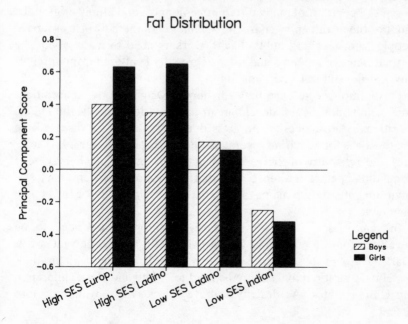

The factors, and percentage of the variance, associated with fat distribution (as contrasted with the amount of fat) were: sex (35 per cent), age (7 per cent), ethnicity (2 per cent), sport (2 per cent), and residual (unspecified) factors are associated with the remaining 54 per cent of the variance.

Malina *et al.* point out that the patterns of fatness and fat distribution of the highly trained Olympic athletes follow the same pattern as that of the general population. In the case of sexual differences, with sex accounting for more than 30 per cent of the variance in fatness and fat distribution, there seems to be a strong genetic determination. Between populations, however, with the percentage of variance due to ethnicity being only two to three per cent, genetic determinants of body composition appear to be relatively weak. Perhaps uninvestigated factors related to the many environmental conditions for growth and development, and their interactions with the genome, that differed between the athletes account for the unexplained variance in fatness and fat distribution.

From this brief review of population variation in growth, it is possible to conclude that the differences between human groups in the average values for size, body proportions, and body composition are due to an interplay between genetic and environmental determinants of growth and development. These variations in the morphology of the human species have their own intrinsic fascination and scholarly appeal for study, but there are also important practical reasons for the analysis of these variations in growth. Cities and nations are becoming increasingly composed of people from many different geographic and ethnic origins. To monitor the health and welfare of the children of these diverse groups of people, clinicians and public health workers need to know about the normal range of amounts and rates of growth of children from different physical and cultural environments.

Furthermore, reference data for height, weight, body composition, and skeletal and dental development are used to assess health status, nutritional status, obesity, progress during treatment for disease, and relative risks for acquiring several acute and chronic diseases. For accuracy and reliability in their work, health care professionals may need population specific reference data that reflect both the hereditary and environmental determinants of growth and development for the people they serve.

In some cases, reference standards for growth may need to be developed for different ethnic groups living in the same country. Garn & Clark (1976) proposed that separate standards of dental maturation, skeletal maturation, and stature be used for white and black children in the United States. As described above, black children mature more rapidly, on average, than white children. This means that black children are often taller than white children of the same age, at least until the whites reach the same developmental stage as the blacks. Thus a white and a black child of the same age, sex, and height may indicate normal growth for the white child and delayed growth for the black child.

In other cases, separate reference standards for growth and body composition may be required for children of similar ethnic origin, but living in different environments. Recall that Johnston *et al.* (1976) found that Europeans living in Guatemala were shorter during childhood than Europeans living in the United States. In other studies, Johnston *et al.* (1975; 1984) and Bogin & Sullivan (1986) found that Europeans living in Guatemala had the same average fatness and fat distribution as native Guatemalan Ladinos of high socioeconomic status, but differed significantly in these measures from Europeans living in the United States. Using North American growth standards, many European children living in Guatemala might be classified as short, underweight, and lean for their age, yet be perfectly normal and healthy by local Guatemalan standards.

Adaptive value of body size in human populations

Implicit in the foregoing discussion of factors that influence population variation in growth is the fact that such variation occurs within the limits of biologically possible human phenotypes. Some of the evolutionary and ecological determinants of the growth and development of the human phenotype were described in Chapters 2 and 3 of this book. Evolution by natural selection is caused by environmental changes, such as variation in food availability and temperature, which test the ability of organisms to adapt to the changes. Adaptation is usually measured by differences in fertility and mortality between individuals of different genotypes and phenotypes, Successful genotypes and phenotypes live, and produce more offspring than unsuccessful types. Cast in this evolutionary framework, the variation in growth, development, and adult size of living human populations may be viewed in terms of its adaptive effect on mortality and fertility. Of course, human beings alive today are the sizes and shapes they are because these have been about the best for their ancestors. Unless past conditions for life still prevail today, the size of living peoples may be unsuited for contemporary conditions. In this case, further adjustments in body size and shape may be evolving.

Widdowson (1968) reviewed experimental studies showing that when fed calorically inadequate diets, non-human animals, genetically bred to have relatively fast rates of growth or large size, suffered greater deprivation than conspecifics bred for slower growth or smaller size. By implication, small body size and slow growth may be adaptive for people living in environments where undernutrition, cold temperatures, or heavy workloads limit the number of calories available for growth.

Garrow & Pike (1967) studied the growth of Jamaican children who had been hospitalized with severe undernutrition, but recovered following nutritional supplementation. Two to eight years after hospital discharge, the subjects were significantly taller, broader in the chest, and had thicker bone and muscle in the leg than their age-matched siblings. It was assumed that rate of growth during recovery was genetically determined, which led the authors to conclude that in this Jamaican population, in which almost all children suffered from protein–energy malnutrition, larger or faster growing infants and children were more likely to experience severe malnutrition than their smaller and slower growing siblings.

Frisancho *et al.* (1973) studied the reproductive outcomes of parents of small and large body size in a Peruvian population. The subjects were living in a shanty-town community, under poor socioeconomic conditions, on the outskirts of a highland city. Infant and child mortality rates of the shanty-town were high, about 286 deaths per 1000 live births, compared with the city as a whole, about 163 deaths per 1000 live births. The percentage offspring survival of the shanty-town parents, calculated as (number of children still alive/number of children born) × 100, was found to differ significantly between short and tall mothers. Women less that 148 cm in height had an average offspring survival of 78.8 per cent compared with 70.9 per cent for women over 148 cm in height.

A study in India by Devi *et al.* (1985) reports very similar findings. The height, weight, and fertility of 291 women living in a coastal fishing village were measured. The villagers were of a fishing caste, belonging to the lower strata of Hindu society. Devi *et al.* described the villagers as undernourished, suffering from numerous infectious diseases, and living with poor sanitary condition and little access to medical care. It was found that maternal stature was negatively, and significantly, correlated with the number of surviving children. That is, shorter mothers (<148.4 cm in height) had an average of 4.39 living children at the time of the study, while taller mothers (<153.6 cm in height) had an average of only 4.01 living children. To the extent that the height of the short stature mothers in both Peru and India was due to genes, rather than nutritional and disease stress during their growth, the difference in offspring survival may be of long-term evolutionary importance. However, Devi *et al.* calculated the intensity of natural selection due to maternal height for their Indian sample and found that it was very low (intensity = 0.08, with 1.00 being the maximum and 0.00 the minimum), indicating that maternal height variation accounted for only a small fraction of the reproductive fitness of this population.

Stini (1975) reviewed several studies of the body size of adults from malnourished communities, including his own research conducted in Colombia, South America. Stini found that in Colombia, and in other countries, adults from undernourished communities were, on average, shorter and had less muscle mass than better nourished groups. According to Stini, muscle mass is a direct indicator of metabolic activity and correlates with the body's requirements for energy and protein. Thus, adults with reduced stature and muscle mass need to eat less total food than people with more skeletal and muscle tissue. This, Stini argued, was a beneficial adaptation to undernutrition, though no direct measures of reproductive success relative to growth reduction were taken. Stini believed that the size reduction associated with malnutrition took place during the prenatal and early postnatal growth of the individual. It was a developmental change, adjusting the growth rate and size of an individual to his environment, and not the result of genetic selection for small body size. Stini argued for developmental plasticity, since 'genetic adaptation alone would frequently result in a stereotypic and potentially maladaptive and rigid response' (p. 35).

In contrast to all of these studies, Martorell *et al.* (1981) found a positive, and significant, correlation between maternal stature and childhood mortality for women living on a coffee plantation in Guatemala. The sample consisted of 380 Indian women, all of whom showed clear evidence of chronic undernutrition. The mean stature of the women was 142.4 cm, with a range from 126.3 cm to 158.6 cm, making them one of the shortest samples of women in the world. Dividing the range of height into terciles, and adjusting the results for the effects of maternal age and parity, it was found that the shortest women had an average of only 2.7 surviving children, while the tallest women averaged 3.2 surviving children. The shortest mothers delivered babies with an average birth weight of 2.85 kg, and the tallest mothers gave birth to babies averaging 3.02 kg. The consequence of natural selection at work in this population is towards favoring taller women who give birth to heavier infants.

Billewicz & McGregor (1982) found differences in mortality according to length and weight for Gambian children between birth and age five years. The children lived in two rural villages, and suffered from chronic undernutrition, often severe during the rainy season, and several infectious and parasitic diseases. The authors used data collected longitudinally from 1951 to 1970. When the data were grouped into one year age samples (e.g., all children from birth to 0.99 years old, 1.0 to 1.99 years old, etc.), it was found that survivors were, on average and in every age grouping, significantly longer and heavier than those who died. In this study, the authors did not correlate child mortality with the size of the mother; however the results do indicate that bigger, and perhaps faster growing, children are better adapted to environmental stress than smaller or slower growing children.

The relationship of adult body size and body composition to fertility in a mildly undernourished community was studied by Mueller (1979). The sample was from highland Colombia, but not the same group measured by Stini (1975). Mueller found that for women, fertility was positively, and significantly, correlated with fatness but not with height or other skeletal dimensions. Since fatness also correlated with income, quality of the home and other indicators of wealth, Mueller concluded that fertility in women was dependent on socioeconomic status, not on physical dimensions. For men, it was found that individuals with both large and small skeletal dimensions had significantly fewer children than men closer to the mean value for stature, biacromial diameter, femur length, etc. Essentially similar results were found by Brush *et al.* (1983) for a sample of 150 women from an undernourished community in highland Papua New Guinea. That is, fatter women had higher fertility than leaner women, and women with the modal height had higher fertility than women significantly taller or shorter than the mean stature. Thus in these two studies there is no evidence of natural selection for smaller body size.

Perhaps these samples of people were not suffering from sufficient environmental stress for selection to occur. Indeed, the findings parallel the results of physique and fertility studies conducted in the United States on healthy, well-nourished populations. Graduates from Harvard University (Damon & Thomas, 1967; Mitton, 1975) and residents of Ann Arbor, Michigan (Clark & Spuhler, 1959; Vetta, 1975) who were closer to the mean value of height, or weight, for their sample had higher fertility than persons closer to the ends of the range in size. In another study, of adults from Boston whose growth during childhood and adolescence had been measured, Scott & Bajema (1982) found that fertility was not associated with stature, but was related to adolescent body proportions. Girls who had a relatively more linear body build, determined by weight/height^2, grew up to become women with significantly higher fertility than girls with a relatively more 'stout' physical appearance. The 'linear' girls were also more likely to marry, which may indicate that the fertility differential was due to social selection for culturally accepted phenotypes, and not natural selection. Finally, Bailey & Garn (1979) studied the relation of physique to fertility for women from the Ten-State Nutrition Study (United States) and found no evidence for natural selection. Rather, a significant socioeconomic effect on both fatness and fertility was evident; poorer women were both fatter and more fertile than wealthier women.

Overall, the evidence for an adaptive value for human body size, other than the modal size of the human species, is tenuous. The hypothesis, derived from non-human animal research, that poor environmental conditions, leading to undernutrition and disease during childhood, may select against genotypes for large size and rapid rate of growth, is not supported by the bulk of human research. To be sure, under conditions of want human populations are, on average, smaller than peoples from less stressed populations. In most cases, however, the smaller average body size of the less fortunate is a result of accommodations in growth and development that occur during the lifetime of the individual, and not evolutionary selection acting across generations. The biological plasticity of the human phenotype (Lasker, 1969) allows the individual to adapt to a very wide range of environmental conditions, and gives the human species an adaptive advantage not found in those species obligated to develop according to a rigid, and predetermined, genetic plan.

5 Environmental factors influencing growth

The previous chapter described population variation in growth, and discussed its causes in relation to the interaction of hereditary and environmental factors. To understand better the contribution of genes, the environment, and their interaction to growth, it is useful to separate artificially these categories and analyze each in turn. The present chapter extends the discussion of environmental factors influencing human growth and development to five comprehensive, and well-documented, categories: nutrition, altitude, climate, migration and urbanization, and socioeconomic status.

Nutrition

In the growing human being, the multiplication of cells or their enlargement in size depends upon an adequate supply of energy, amino acids, water, lipids, vitamins, and minerals. Growth and nutrition are, therefore, closely correlated. People require approximately 48 essential nutrients for growth (Guthrie, 1986). Essential nutrients are defined as those that people cannot produce naturally from simpler elements, and if eliminated from the diet of an otherwise healthy, well-fed individual will result in growth failure. The study of human growth and nutrition is hampered by the fact that direct controlled experimentation in the area of nutrient deprivation is unethical. To define the types and amounts of nutrients essential for human growth nutritionists rely instead on experimentation with non-human animals, on the study of normal humans, on the analysis of naturally occurring human malnutrition, and on the response to nutrient supplementation of people suffering from malnutrition.

Adequacy of the total quantity of food consumed is a major determinant of growth. In populations where food shortages are present, growth delays occur, and children are shorter and lighter than in populations with adequate or overabundant supplies of food. The famines of the World Wars of 1914–18 and 1939–45 retarded the growth of children and adolescents exposed to them (Wolff, 1935; Howe & Schiller, 1952; Markowitz, 1955; Kimura & Kitano, 1959). For example, in a review of Japanese studies Kimura (1984) found that mean stature of children

126

decreased between 1939 and 1949, and returned to pre-war levels in 1953 for girls and 1956 for boys. The growth of children who were between the ages of birth and 12 years during the war was affected more than the growth of older children (showing, once again, how pre-adolescent growth is more sensitive to the environment than adolescent growth). The post-war recovery in height and weight 'occurred most rapidly in large cities, followed by small cities, and then rural mountain villages' (p. 200). Since improvements in diet also followed the same path, it is likely that the growth recovery was, in large part, due to better nutrition.

Populations that depend on subsistence agriculture for their food may face periodic food shortages due to variation in rainfall, temperature, crop diseases and pests, and inadequate food storage. Billewicz & McGregor (1982) analyzed the growth of children and adults living in two Gambian (West Africa) villages. Longitudinal measurements of height and weight, along with extensive medical histories, were recorded in these villages since 1951. The authors described agricultural practices as 'primitive.' Vegetable foods were the main source of subsistence, though meat was eaten on religious festival days. The agricultural cycle was determined by climate. There was a dry season and a rainy season, the latter lasting from late May to the end of October, with August and September being the wettest months. Food supplies were lowest from August to November. Typically, adults lost 2.5 kg in body weight during the rainy season. The mean weight of 157 men over 25 years old was 59.8, 56.8, and 58.8 kg in March 1966, November 1966, and March 1967. For a sample of 201 women the corresponding mean weights were 52.6, 50.4, and 53.1 kg. Children grew significantly faster in height and weight during the dry season than during the rainy season. For boys and girls aged five to nine years old, the dry season increase in height averaged 6.1 cm and the rainy season increase averaged 4.2 cm. Food shortages during the rainy season occurred simultaneously with an increase in the incidence of malaria, intestinal parasites, and childhood gastroenteritis. These diseases may decrease the food intake of affected individuals, increase protein and energy expenditures to combat the disease, or decrease the nutrient value of foods consumed (due to parasite competition, diarrhea, etc.). The combination of these insults resulted in severe undernutrition during the rainy season, directly reflected in the poor growth of children and the weight losses of adults. Billewicz & McGregor found that the growth deficits associated with rainy season undernutrition were not overcome during the dry season, and that over the year the people suffered from a net shortage of calories. As a result, the Gambian children grew less at every age compared with a reference sample of British children. For instance, at three months of age Gambian children were 1.0 cm shorter than British children, but by three years of age the difference averaged 6.9 cm, and this difference was maintained to adulthood. The adolescent growth spurt of Gambian boys was delayed (see Figure 4.1), peak height velocity took place at 16.3 years for the boys compared with about 14.2 years for British children, and the intensity of the spurt was reduced, peak height velocity was 6.9 cm/yr compared with 8.2 cm/yr for British children. The Gambian children had a longer total growth period, which made up some of the difference in height, but the net result of a lifetime of undernutrition was a significant reduction in height and weight compared with better nourished populations.

Populations that have not experienced the acute starvation of wartime famine, or seasonal food shortages associated with subsistence agriculture, may still show a strong relationship between malnutrition and growth. Findings of the Ten-State Nutrition Survey, conducted in the United States from 1968 to 1970, were reviewed by Garn *et al.* (1974), Garn & Clark (1975), and Lowe *et al.* (1975). Anthropometric, dietary, and biochemical indicators of nutritional status were collected from 40 847 individuals of white, black, and Hispanic background. Most of the sample were low income families, with the lowest incomes disproportionately represented. Children from lower income families were shorter and lighter than children from higher income families, and this was true across all levels of income and within each ethnic group. Children from lower income families consumed significantly less total food, and thus received fewer calories from all sources, than children from higher income families. Relative percentages of nutrients were about equal in the diet of all economic and ethnic groups, except for vitamin C, which was more abundant in the diet, and in the blood serum, of higher income families. The differences between income groups in stature and weight were significantly correlated with the difference in caloric consumption. Garn & Clark (1975) pointed out that the poor in the United States were not suffering from acute undernutrition, but the moderate, chronic undernutrition they did experience resulted in a cumulative growth deficit.

Jenkins (1981) studied the growth and nutritional status of 750 children belonging to four ethnic groups living in Belize. The groups were Creole, Garifuna, Mestizo, and Maya. Creole and Garifuna are of Afro-Caribbean descent, although it is popularly believed that Creoles have a significant amount of European admixture. Both groups speak English and follow Black Carib cultural practices. Mestizos are of mixed Spanish–Indian descent, speak Spanish, and follow Latin American cultural practices. The Maya are descended from native Indian peoples, they speak languages of their local culture (Mopam and Kekchi), and follow traditional Indian cultural practices. Within Belizean society, Creoles and Mestizos have higher socioeconomic status, based on parental education and occupation, than the Garifuna or Maya, although all groups are of lower socioeconomic status compared with white Europeans living in the country.

Jenkins found that Maya and Garifuna children were generally smaller and more frequently malnourished than Mestizo or Creole children. Poor growth and nutritional status were associated with the frequency and severity of diarrhea, a later age of introduction to solid food, and a larger number of children in the household. Episodes of diarrhea probably reflected the exposure of children to infectious disease and contaminated food and water. Age of introduction to solid food reflected cultural differences in infant feeding practices. Number of children is a measure of family size, and, since larger families have more 'mouths to feed,' there is usually a negative correlation between family size and the growth and nutritional status of children. Jenkins concluded that ethnic variation in the growth of Belizean children reflects 'differential access to medical, sanitational, and dietary resources' (p. 177) of the groups studied. Access is mediated by cultural practices and economic resources, and of these, family economics was the most important; undernutrition was nearly always associated with poverty.

Bailey *et al.* (1984) studied the growth of more than 1000 children living in 29 villages in northern Thailand. The children lived in rural agricultural villages, with rice as the basic subsistence crop. The villages had schools and received some type of health care, but none had electricity and only one was located near a major highway. The children were between the ages of six months and five years old at the start of the study, and they were measured about every six months for five years. Children born during the study were also measured. This study design provided cross-sectional and longitudinal data on recumbent length (at all ages), weight, head circumference, skinfolds, hand–wrist radiographs, nutritional biochemistry from blood samples,and parasite infestations. Disease histories were collected on each child, every 15 days, for three years of the study. Some of the children were given nutritional supplements (amino acids, thiamine, vitamin A, and iron) to assess the effect of supplementation on growth.

It was found that, compared with local or international reference standards for growth, these rural Thai children were delayed in growth of all the dimensions studied. For instance, compared with a sample of healthy middle class children from Bangkok, the capital city of Thailand, the rural village children were on average about 4.7 cm shorter, 1.3 kg lighter, and 1.2 cm smaller in head circumference over the first 36 months of life. From ages six to 18 months, the rural Thai children were between the fifth and 10th percentiles of length and weight of a national sample of United States children. The growth of Thai children older than 18 months fell to below the third percentile of the American data. When nine-year-old children were compared with local and international reference data, skeletal maturation, as assessed from the hand–wrist radiographs, was delayed up to 34 months in girls and up to 13 months in boys.

Disease histories, parasite infestations, and mortality during childhood were not significantly associated with growth in this sample. Children receiving the nutritional supplement did not grow significantly larger or faster than did children on the unsupplemented diet. Bailey *et al.* concluded that the delays in growth were not due to disease or the lack of specific nutrients, such as protein, vitamin A, or iron; rather the delays were due to a deficiency in the total intake of calories. The falloff in growth at 18 months of age corresponded with the average age at weaning in these villages. Weaning foods were usually watered-down versions of adult foods. Although there were no food shortages in the villages, the small gastrointestinal tracts of the weaned infants and young children may not have been capable of digesting enough food to meet their caloric demands for maintenance and growth of the body.

Behar (1977) found that the same problem was faced by young children in rural villages in Guatemala. Young children had access to sufficient food, but the traditional diet of corn and beans did not have the caloric density to meet their growth requirements. Waterlow & Payne (1975) reviewed several studies that weighed and analyzed the food intakes on one- to two-year old children not receiving any breast milk. Results of the analysis for protein and energy (calories from all sources) content of the food is presented in Figure 5.1, in which the percentages of the recommended daily allowance, according to United Nations guidelines, are compared. Protein intakes are close to or exceed 100 per cent of the recommended allowance in every country, but energy intakes are below recommendations in every case. These data support the conclusions of Bailey *et al.* (1984) and Behar (1977) that calorie insufficiency, due to feeding foods inappropriate for the digestive capabilities of young children, leads to undernutrition and growth delay.

This research indicates that cultural behaviour related to infant and child feeding may determine differences between populations in nutritional status and growth. One specific example is a study by Jenkins *et al.* (1984) of the feeding practices and growth of infants and children of the Amele culture of Papua New Guinea. Amele parents classify several stages of child growth according to developmental landmarks, and ascribe to each stage appropriate foods. Newborns may only be fed breast milk or warm fruit and vegetable juices. Amele mothers believe that the quality of breastmilk improves as the infant grows older, so breast-feeding is often prolonged up to four or five years. Infants who can sit alone may be given some mashed foods mixed with liquids, but not roasted foods, or those mixed with coconut milk, as these foods are too dry or slimy and may induce coughs. By the time a child can run and play with other children he may request and be given all types of food.

Jenkins *et al.* grouped 22 infants, for whom there were longitudinal growth data from birth to 24 months, according to the Amele developmental stages and calculated the percentage of children gaining, maintaining, or losing weight relative to international standards of weight for age. During the first growth stage, roughly from birth to six months, 77 per cent of the infants gained weight relative to the standard. Thus, breastmilk seemed adequate for the majority of these infants. During the next stage, from about six to 12 months, 82 per cent of the infants lost weight on the predominantly breastmilk dependent diet; the fruit and vegetable juice supplements were of little caloric or other nutritional value, since most of these supplements were more than 90 per cent water. Breastmilk and these liquid supplements did not meet the growth needs of the infants. The next two growth stages cover the period from roughly 12 to 24 months. As children began to sit, crawl, stand, and walk, progressively more solid foods of plant and animal origin were added to

Figure 5.1. Percentage of United Nations (Food and Agricultural Organization and World Health Organization) recommended intakes of protein and energy of weaned children, one to two years old, in five developing nations. From Waterlow & Payne (1975) and Johnston (1980).

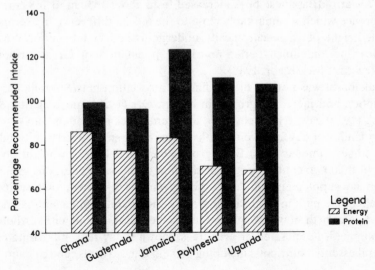

the diet. The percentage of chidren losing weight relative to the standard fell to 28 per cent in the third stage and about 15 per cent in the fourth stage. Although Amele children age 24 to 36 months appeared to grow adequately, they achieved heights and weights at only the fifth to 10th percentile of the international standard. Thus they were relatively small as children, and measurements of adults show that they remain small.

Another example of how culture, food, and growth are related was described by Takahashi (1984), who found an association between changes in dietary practices and growth in Japan. Rice has been, and still is, the dietary staple of Japan. Until 1950, fish and shellfish were the major sources of animal protein, although most dietary protein was of plant origin, from soybean products. Post-war changes in Japan, including greater contact with Western cultures and economic development, altered the traditional diet. These diet changes began in the late 1950s, but became pronounced in the mid-1960s. From 1966 to 1976, rice consumption decreased from about 350 to 225 grams per person per day. During the same time, meat consumption rose from about 35 to 60 grams per person per day and milk consumption rose from about 55 to 100 grams per person per day. The height of school boys, aged six to 17 years, rose by an average of 4.1 cm between 1930 and 1960, a period of relatively great social and economic change, but rose by an average of 5.3 cm between 1960 and 1975. Takahashi attributes almost all of the increase in the 1960 to 1975 period to changes in diet, especially the increased consumption of milk. Other factors, such as lower rates of childhood disease and reduced family size, may also have contributed to the height increase, but the result is that between 1960 and 1975 the average height of 17 year old Japanese boys increased from about 163 cm to 168 cm, a difference which is remarkably close to the 6.4 cm difference in average adult height between chronically undernourished Gambians of West Africa and the much better nourished population of Great Britain (Billewicz & McGregor, 1982).

Takahashi was not the first to find an association between milk consumption and increased growth in height. Orr (1928) and Leighton & Clark (1929) gave school children, in several cities in Scotland, an extra pint of milk per day for seven months. Some children received whole milk and some skimmed milk. Both groups increased faster in height and weight than two control groups of children of the same ages, one group given no supplement and the other given a supplement of biscuits, equalling the milk in total calories. Since the biscuit supplement had no effect on growth, it was concluded that some factor in milk, either whole or skimmed, accelerated growth. It is known today that milk contains several essential nutrients, including protein (amino acids) and calcium.

To further support the 'milk hypothesis,' Takahashi (1984) reviewed data on the growth of pastoralists, living in traditional, non-Western cultures, whose diet is based on animal milk, and agriculturalists, whose diet is usually devoid of milk and milk products. The pastoralists of Central Asia (peoples of the Gobi, Takola Makan, and Kavil Deserts) and the pastoralists of East Africa (the Masai, Samburu, and Datoga) were found to be taller than their rice or grain growing counterparts.

Growth of the Turkana, a pastoral people of Kenya, East Africa, was studied by Little *et al.* (1983). Their findings support Takahashi's 'milk hypothesis.' The basic staple of the Turkana diet is milk, supplemented with blood and meat from their animals and grains and sugar which are acquired through trade. Little *et al.* found that the Turkana are significantly taller than their agricultural neighbors with whom they trade. As may be seen in Table 4.1, at age 20 years the Turkana are shorter than Americans. Little *et al.* found that the Turkana continued growing until aged 23 years, at which time they are as tall, on average, as Americans. The mean weight of the Turkana is significantly below that of American blacks and whites at all ages. Little *et al.* suggested that the Turkana have high protein, but low calorie intake that 'contributes to an adequate lean tissue disposition but limits the storage of adipose tissue' (p. 826). Thus, the energy supplied by the milk and other foods of animal origin is not the factor that results in the increased growth in height. Whether this factor is protein, calcium, iron, or a combination of these and other nutrients is not known.

By definition, the lack of specific essential nutrients, such as those supplied by milk, can delay growth and may be the cause of some population differences in size. Greene (1973) studied the growth of Ecuadorian Indians living in an area with chronic iodine deficiency. Men without clinical signs of iodine deficiency (goiter, cretinism, etc.) averaged 155.7 cm in height. Male deaf–mute 'cretins' averaged 146.2 cm. The difference is statistically significant, and a similar difference existed between normal and affected women. The heights of normal men and women in the study area, and in two other areas in the 'low iodine' zone of Ecuador, were compared with the heights of Quechua Indians from Peru. The Ecuadorian men averaged 155.5 cm and the Ecuadorian women averaged 144.8 cm. The Peruvian men and women averaged 160.0 cm and 148.0 cm in height. All the individuals were of Indian ethnicity, of low socioeconomic status, and lived at high altitude. Greene argued that the major environmental difference between the Ecuador and Peru samples, accounting for the difference in stature, was the low iodine availability in Ecuador.

Although the deficiency of any essential nutrient will result in growth retardation, these kinds of nutritional problems are rare in comparison with mild-to-moderate energy deficiency, which is ubiquitous in the lower socioeconomic status populations of the developing nations of Africa, Asia, and Latin America and also present among the disadvantaged peoples of the developed nations.

The biological response of children to calorie undernutrition is to delay growth in the skeleton and in the soft tissues of the body and to delay the rate of maturation towards adulthood. Bailey *et al.* (1984) reported that delays in skeletal maturation allowed undernourished rural Thai children to grow for, on average, at least a year longer than United States children. However, the delays in length growth that the Thais experienced over their lives meant that, as adults, they reached only the fifth percentile for height of the United States reference population. Mild-to-moderately undernourished children in rural Peru (Frisancho *et al.*, 1970) and Guatemala city (Bogin & MacVean, 1983) were found to be shorter and delayed in skeletal maturation compared with better nourished children. In both cases, the relative difference in stature between the malnourished and better nourished samples was greater than the relative delay in skeletal maturation. This suggests that, like the Thai children, the undernourished Peruvians and Guatemalans would also be short statured as adults.

A similar pattern of growth was found for rural Indian boys by Satyanarayana *et al.* (1980). Boys who had been malnourished and delayed in growth at age five years entered adolescence about two years later than better nourished boys. Despite their longer period of growth, the malnourished boys were significantly shorter than the better nourished boys at age 18. At age five the two groups differed by 16.5 cm in height (105.0 cm versus 88.5 cm), and at age 18 the groups differed by 15.5 cm in height (164.5 cm versus 149.0 cm). Though the malnourished boys were likely to continue growing for about two years longer (and adding perhaps 5 cm in stature) than the better nourished group, the significant difference in height would persist into adulthood.

Altitude

More than 25 million people live in high altitude regions of the world, that is, at altitudes of 3000 meters above sea level or higher (Baker, 1977). The effects of high altitude on the growth of populations living in the Andes Mountains (South America), the high plains of Ethiopia (Africa), and the Himalaya Mountains and Tibetan Plateau (Asia) have been studied most intensively. High altitude environments impose a number of stresses on people, including hypoxia, high solar radiation, cold, low humidity, high winds, and rough terrain with severe limitations of agricultural productivity of the land. Of these, hypoxia, the lack of sufficient oxygen delivery to the tissues of the body, is the most severe since it cannot be overcome by any cultural or behavioral adaptation available to native high altitude peoples. The cause of hypoxia is the low partial pressure of oxygen in the atmosphere above 3000 meters. The partial pressure of oxygen in the air and in the alveoli of the lung, the site of oxygen exchange between the lung and the red blood cells, at altitudes from sea level to 15 000 meters is shown in Figure 5.2. The partial pressure of oxygen in the alveoli determines, in part, the saturation of hemoglobin with oxygen. Red blood cells carry the oxygen saturated hemoglobin from the lung to the tissues of the body where the oxygen is used to maintain normal metabolic activity. At sea level, the hemoglobin of the red blood cell is about 97 per cent saturated with oxygen as it leaves the lung, but at 3000 meters the arterial hemoglobin is only about 90 per cent saturated with oxygen. The decrease of oxygen saturation is sufficient to disrupt cellular metabolism (Luft, 1972), and may delay cell growth.

In humans, prenatal growth retardation at high altitude, as evidenced by low birth weight, was shown by Lichty *et al.* (1957) for infants born in Leadville, Colorado (USA), altitude 3000 meters, and by Hass *et al.*

Figure 5.2. The relationship of altitude above sea level to the partial pressure of oxygen in the atmosphere and in the alvioli of the lungs.

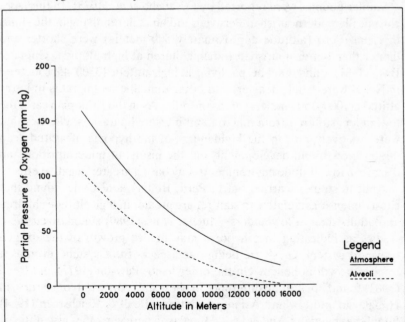

(1980) for infants born in La Paz, Bolivia, altitude 3600 meters, compared with lowland births from the same populations. Hass *et al.* controlled for the effects of maternal nutritional status and smoking, gestation length, ethnic background (Indian or Mestizo), and length of residence at high altitude. When statistically adjusted for these confounding variables the mean birth weight of low altitude infants was 3415 grams and that for high altitude infants was 3133 grams, which is a statistically significant difference. The mean crown–heel length of the low altitude sample was 49.6 cm and that for the high altitude infants was 49.0 cm, also a significant difference. Thus, prenatal growth of the skeleton, as well as soft tissues, is reduced at high altitude.

Frisancho & Baker (1970) found that Peruvian Indian children living above 3000 meters were shorter and lighter, on average, than lowland Peruvian children of the same age. Similar findings were reported by Beall *et al.* (1977) and Mueller *et al.* (1978; 1980). Frisancho & Baker (1970) characterized the high altitude children as having a slow rate of growth, a prolonged growth period, lasting to age 22 years, and a later and poorly defined growth spurt. In contrast, Clegg *et al.* (1972) found that high altitude living Ethiopians were taller and heavier than ethnically similar people living at low altitude in Ethiopia. In this case, the lowland population suffered from malaria and intestinal parasites that may have compromised their growth.

Similar findings were reported by Frisancho *et al.* (1975). In this case, chronically undernourished Peruvian Indian children living in the slums of Lima, Peru (altitude approximately 500 meters) were shorter and lighter than better nourished Indian children at high altitude. Gupta & Basu (1981) found that Sherpas living at high altitude (3500–4500 meters) in Nepal were taller, heavier, and fatter than Sherpa migrants to lower altitude (1000–1500 meters) areas of India. As in the Ethiopian case, the lowlanders suffered from malaria and intestinal parasites, diseases that were relatively rare in the highlands. So, the hypoxia of altitude can inhibit growth and development, but the insults of undernutrition and disease in low altitude environments may have a greater negative effect.

Andean studies, carried out in Peru, Bolivia, and Chile, found that chest dimensions, relative to stature, are greater in high altitude children and adults than in lowlanders. Mueller *et al.* (1980) summarized these studies as indicating that hypoxia may induce growth of the oxygen transport system, chest size being one aspect of this system, even as it retards growth in height. On the other hand, Pawson (1977) and Beall (1982) found no altitude difference in chest size, corrected for stature, in Himalayan and lowland Asian populations. Beall & Reichsman (1984) found that highland Andean and Himalayan native peoples also differ in blood biochemistry; Andeans have higher hemoglobin concentrations than Himalayans or lowlanders. This suggests that the ability of Andeans to reside successfully at high altitude depends on adaptation to hypoxia at the morphological and vascular level (e.g., chest size, lung volume, and hemoglobin concentration), while the Himalayan adaptations may be at the tissue level (e.g., more efficient cellular utilization of available oxygen). If true, this would account for differences in chest growth, and some other skeletal dimensions, between populations of the Andes and Himalayas.

Perhaps the clearest evaluation of the effect of hypoxia on growth comes from the work of Stinson (1982). She measured 323 children, between the ages of eight and 14 years, attending a private school in La Paz, Bolivia. The children, mostly of Spanish or Spanish–Indian ancestry, were of middle to upper socioeconomic class, healthy, and well nourished. The children had resided at high altitude for various amounts of time. Stinson found a negative correlation between stature and length of residence in La Paz, that is, the shortest children had lived at high altitude the longest time. The altitude effect was small, however, with about a 2.5 cm difference in height between children with the longest and shortest residence at high altitude. In contrast, Frisancho (1977) found an average difference in height of five to eight centimeters between highland and lowland Mestizo children in Peru. In Peru, the difference in height was due to high altitude hypoxia, nutritional status, disease rates, and socioeconomic conditions. Each of these variables interacts with others, and confounds the independent influence of hypoxic stress on growth.

Climate

Heat, cold, and relative humidity are associated with variation in the size, proportions, and composition of the human body. Roberts (1953) found a significant negative correlation between body weight and mean annual temperature for a world-wide sample of populations. He noted also that the ratio of sitting height to total height decreased as temperature increased. When Roberts separated the world sample into geographic groups, he found these same temperature dependent relationships within European, African, Asian, and Amerindian samples. Taken together, these results show that in colder climates people tend to be heavier with relatively larger trunks and shorter legs, while in hotter climates people tend to be lighter and relatively longer legged.

Newman (1953) found similar results for a large sample of Amerindian populations of North America, but within the more restricted geographic range of these populations the climate effect is much smaller than that found by Roberts. Froment & Hiernaux (1984) surveyed sub-Saharan

African populations and found that body length measurements, such as stature, correlate positively with mean temperature of the hottest month and negatively with humidity of the driest month. That is, Africans living in hot–dry areas tend to be taller than Africans living in hot–wet areas. Schreider (1964a) used height and weight measurements from samples of people living at all latitudes to show that body surface area increases, on average, from colder to hotter climates.

These findings conform to ecological 'rules' of mammalian biological adaptation to the thermal environment. In hot environments, excess body heat produced by mammalian metabolism and voluntary muscular activity must be dissipated to the environment to avoid hyperthermic stress. Such loss may occur by radiation (direct transfer of infrared energy from the body to a cooler object), conduction (heat exchange by direct physical contact between the body and a cooler object), convection (heat exchange between the body and a cooler object via an intermediary medium, e.g., air flow), or evaporation (conversion of water, e.g., perspiration, to vapor using body heat). Relatively low body weight, or body volume, and relatively large body surface area, produced by having legs and arms relatively long in proportion to the size of the trunk of the body, assist in heat loss. Low body volume decreases the distance required for the radiation of heat from the internal organs and muscles to the surface of the body. Large body surface area increases the potential for convection, conduction, and evaporation. In cold environments, a relatively large body volume and small surface area (i.e., relatively short extremities in proportion to trunk size) is the body type best suited for heat retention. Body fatness, especially the thickness of the subcutaneous fat layer, may also increase in cold environments. Adipose tissue is relatively inert metabolically, owing to poor vascularization, and can act as an insulating barrier against radiative heat loss. In hot environments, a thin subcutaneous layer of fat would help minimize heat retention.

The studies cited above, as well as more recent work, show that these rules are generally followed by human populations. Hiernaux (1974) showed that the small stature and lean body build of the Mbuti pygmies and other pygmoid peoples of Africa is adapted to the tropical rainforest. The most severe stress of the rainforest is the very high humidity of the rainy season, which, combined with moderately high temperatures, limits heat loss from the body by evaporation. Radiation and convection must be maximized in this situation, and the small body mass of the pygmy peoples maximizes the avenue for radiative heat loss. Populations inhabiting the Amazonian rainforest also are short statured and have small body mass. In some cases, such as the Maku people of Brazil and Colombia studied by Milton (1983), caloric undernutrition may be the cause of small body size. But, in a survey of other Amazonian cultures, Milton did not find evidence for undernutrition. Rather, the evidence favored the thermoregulatory value of small body mass in tropical rainforest adaptation.

Crognier (1981) surveyed 85 European, North African, and Middle Eastern populations and compared eight climatic variables with 14 anthropometric measurements. Crognier hypothesized that mean annual low temperature would be the strongest influence on the body size and body proportions of the populations studied. The results showed this was true for most cranial measurements, but the post-cranial measurements had their strongest correlation with heat and dryness. Crognier argued that effective cultural adaptations to cold, such as fire and clothing, have been in use since human groups first occupied the temperate latitudes. These means of combating cold may have relaxed selection pressures on biological adaptations to low temperature. In contrast, cultural adaptations to heat are, perhaps, less biologically effective than cultural adaptations to cold. This, Crognier believes, may explain why the populations surveyed showed the association between heat and post-cranial anthropometry. Crognier suggested a biological explanation for the correlation between cold and cranial measurements. He argued that the brachycephalic (rounder) head shape found in the populations studied is better suited for cold than the dolichocephalic (longer) head shape, since a sphere affords maximum volume for heat retention and minimum surface area for heat loss.

This 'ecological' argument for head shape, and other similar arguments for a biological function for head shape, is not very compelling. The same cultural means of protecting the body from cold could be used to protect the head, and are used by Eskimos, who are about as dolichocephalic as Africans. Furthermore, other cultural practices, such as normative sleeping positions for infants and children, can alter head shape during growth. Swaddling practices that force infants to sleep lying on the back produce rounder skulls; sleeping with the head turned to one side produces longer skulls. Boas (1912) showed that migration from central Europe to New York, and other northern American cities, changed the shape of the migrants' skulls from the brachycephalic shape of the parents to a more dolichocephalic shape of their children in one generation. Given these caveats, there is little support for an adaptive or evolutionary explanation for head shape in the populations studies by Crognier, or in any other population.

Climatic variation can also influence growth rates during the year. Buffon (1777) published the first data suggesting the existence of seasonal variation in the growth rates of healthy children. Studying the growth records of the son of Count Montbeillard, Buffon noted that most of the boy's height increase during the year took place in the summer months. Since Buffon's time, dozens of investigations of seasonal growth, using larger sample sizes, have generally confirmed that at temperate latitudes, healthy well-nourished children grow more quickly in height during the spring and summer than they do during the fall and winter (Bogin, 1977; 1978). Studies of seasonal variation in weight gain and loss found that for healthy individuals, fall or winter are the seasons of maximum weight gain and summer tends to be the season of minimum weight gain for children, and the time of maximum weight loss for adults. Several researchers also found that minimum weight gains, or even weight losses, occurred simultaneously with maximum height gains (Bogin, 1979).

The cause of seasonal variation in height growth rate is not completely understood; however, several investigators have suggested that seasonal periodicity in sunlight may act on the human endocrine system so as to synchronize changes in growth-regulating hormone activity with changes in sunlight availability or intensity. Nylin (1929) experimentally tested this hypothesis on a sample of Swedish boys. He exposed one group of boys ($n = 45$) to 'sunlamp' treatments (using a lamp that produced both visible and ultraviolet light) during the winter months in Stockholm and compared their growth to a group of boys ($n = 292$) not receiving treatment. Situated at about 59° N latitude, Stockholm has only six hours of daylight at winter solstice. During the period of treatment, the experimental group averaged 1.5 cm more growth in height than the control group. During the summer, the control group grew at a faster rate than the experimental group, so that over the entire year there was no difference between groups in total height gain. Only the time of year when the maximum gain occurred differed between groups. Marshall & Swan (1971) compared the monthly growth rates of blind and normally sighted children living in southern England over a period of a year. Both groups showed rhythmic changes in growth rate of equal magnitude. The months of maximum growth of individual blind children were distributed evenly throughout the year. The maximum growth rates of the normally sighted children virtually all occurred between the months of January and June. The authors suggested that seasonal variation in day length might have affected the growth rates of normally sighted children. To investigate further this hypothesis, Marshall (1975) analyzed monthly measurements of height, taken over a two year period, for 300 healthy children living on the Orkney Islands (59° N latitude) and concurrent changes in day length, hours of insolation (hours during the day when the sun is not obscured by clouds or haze), rainfall, and temperature. No significant associations were found between monthly increments of height and any of the meteorological variables. Marshall concluded that there was no support for the 'day length' hypothesis.

However, two other studies lend support to an association between variation in sunlight and growth rate. Vincent & Dierickx (1960) found that healthy children living near the equator in Kinshasa (then Leopoldville), Zaire grew more rapidly in height in the dry season than in the rainy season. Diet, temperature, humidity, and sunlight variation were considered as possible influences. No evidence in favor of the first three was found. Although day length was slightly longer during the rainy season, there were far more hours of insolation (bright sunshine), and more opportunity for children to be exposed to the sun, during the dry season. The authors concluded that exposure to sunlight could regulate growth rate. Bogin (1978) measured monthly increments of height growth over a 14 month period for a sample of 246 healthy children of high socioeconomic status living in Guatemala City (14° N latitude). The children aged five to seven years old grew at a faster rate during the dry season than during the rainy season. As in the African study, day length was longer during the rainy season, by about 1.5 hours, but there were significantly more hours of insolation (direct sunshine, without clouds or haze blocking solar radiation) during the dry season (1662 hours from October through April) than during the rainy season (962.8 hours from May through September). Thus, these two studies show that the strongest association between sunlight and growth rate was not for day length, but for hours of insolation.

A relation between light and growth has been known since 1919, when it was shown that ultraviolet light could cure rickets, a disease of bone growth. A few years later the relationship between vitamin D and normal bone growth was demonstrated (see review by Cousins & Deluca, 1972). It is now known that cholecalciferol, vitamin D_3, is synthesized by the skin (and modified in the liver and kidneys to its active form) when people are exposed to ultraviolet light. The physiological action of vitamin D_3 is to increase the intestinal absorption of calcium and to control the rate of skeletal remodeling and the mineralization of new bone tissue (Rasmusen, 1974; Vaughn, 1975). Vitamin D_3 is essential for normal bone growth, and thus growth in height.

Vitamin D_2, the form added to milk and other foods, has these same general properties, but is many times less potent in action (Wurtman, 1975). Haddad & Hahn (1973) working in St Louis, Missouri and Stamp & Round (1974) working in London, England found no correlation between dietary intake of vitamin D_2 and levels of vitamin D_2 metabolites in the circulatory system of healthy white adult subjects. Both investigations did find significant month to month variations in the circulatory levels of vitamin D_3 metabolites, and the variations were associated with the availability of ultraviolet radiation from the sun. These studies show that the major source of vitamin D is the body's synthesis of the D_3 form, and not dietary supplements of the D_2 form. In both St Louis and London the amount of ultraviolet light penetrating the atmosphere and reaching the ground at noon is 15 times greater in June than in December (Wurtman, 1975). Variations in day length, outdoor exposure, and clothing requirements increase the difference between seasons in the exposure of people to ultraviolet light. Although no direct studies have been made, it seems that variations in ultraviolet light and vitamin D_3 synthesis may be related to the general finding that children living at temperate latitudes grow faster in height in the spring and summer and slower during the fall and winter.

In the tropics, where variation in day length is minimal, the amount of ultraviolet light reaching the ground is a function of the amount of cloud cover blocking the sun's radiation. Physicists call the process whereby light energy is reduced when passing through any medium 'extinction.' The extinction of ultraviolet radiation occurs when sunlight entering a cloud is scattered by collisions with water droplets or suspended particles. It is estimated that a thick white cloud will extinguish 90 per cent of the ultraviolet light entering it (Van de Hulst, 1957). Thus, at tropical latitudes, the measurement of hours of insolation provides a fairly accurate indicator of available ultraviolet light and rate of vitamin D_3 synthesis in the body. The studies of seasonal variation in growth conducted in Zaire and Guatemala found that the availability of ultraviolet light, and the opportunity to be exposed to it, are greater in the dry season than during the rainy season. Greater amounts and exposure to ultraviolet light may have increased the rate of vitamin D_3 synthesis, calcium absorption, skeletal remodeling, bone mineralization, and growth in height of the children measured in these two studies.

The causes of seasonal variation in the rate of growth in weight seem to fall into two categories, depending on the circumstances of the populations studied. The first category includes populations suffering from seasonal food shortages or disease, such as the people of the Gambia studied by Billewicz (1967) and Billewicz & McGregor (1982). In these cases, the cause of weight growth variation is clear.

The second category includes healthy, well-nourished populations, and in this case the causes are not so easily discerned. Bogin (1977) reviewed the literature relating to seasonal variation in the rate of growth in height and weight. He found that 22 of the 29 studies reviewed showed that maximum increments in height and weight do not occur at the same time of the year. The authors of some of these studies speculated that children have a natural, endogenous rhythm for growth in weight that is independent of the seasonal rhythm of growth in height.

In an original study, Bogin (1979) measured monthly increments of weight change of 85 healthy, well-nourished children (43 boys and 42 girls) living in Guatemala, and found that minimum weight gains occurred for most individuals (90 per cent of the boys and 80 per cent of the girls) during the dry season, the time of maximum increments in height. In the last three months of the dry season, between 30 and 60 per cent of the children did not gain weight, or even lost weight, although they were growing in height. The pattern of weight change could not be explained by observable behavior related to diet, exercise, or disease. Perhaps the change in metabolism that produced the maximum increments in height growth resulted in a loss of soft tissue. Unfortunately, direct measures of soft tissue were not taken.

Other problems associated with the study of weight changes in healthy, high socioeconomic status populations, are that voluntary 'dieting,' related to seasonal trends in personal fashion, and periodic changes in energy expenditure, due to seasonal leisure activities, influence weight changes. Studies that rigorously control for each of these confounding factors are needed to ascertain the presence or absence of meteorologically produced seasonal changes in weight growth.

Migration and urbanization

Meredith (1979) and Malina *et al.* (1981) reviewed studies from the period 1880–1920 that showed rural living children in the United States and Europe were taller than their urban peers. For example, Steegmann (1985) found, in a study of 18th century British military records, that the stature of rural born recruits averaged 168.6 cm and the urban born recruits averaged 167.5 cm, a statistically significant difference. By 1930 this pattern was reversed and urban children were consistently taller and heavier than rural children. Other sources of data (Tanner & Eveleth, 1976; Fogel, 1986; Bogin, 1988) provide some ideas about changes in the rural and urban environment at the turn of the century. Prior to 1850, mortality rates for infants, children, and adults were higher in the cities than in the rural areas of Europe and the United States. Around 1900 the trend reversed, owing to improvements in sanitation, water treatment, and food preservation in the city. Urban children benefited from these improvements, as reflected in their greater stature and weight, and more rapid maturation, compared with rural children.

Throughout recorded history, social, economic, political, and biological events encouraged the development of cities and the migration of rural people to urban areas. For instance, opportunities for wage-earning employment, education, and health care have been more available, generally, in the city than in the countryside. Given this, it is not surprising that rural-to-urban migration is the most common type of migration that has occurred during recorded history (Smith, 1984), and this type of human migration is occurring more rapidly today than ever before, especially in the least economically developed nations of the world (United Nations, 1980). Demographic change during the past 200 years provides a glimpse of the extent of urban migration. In the year 1800, there were about 25 million people living in urban areas. In 1980, there were about 1.8 billion and by the year 2000 it is estimated that there will be 3.2 billion, which is a 128 fold increase in two centuries. By contrast, the natural increase of the world's population will be only 6.4 fold in the same two centuries (about one billion people in 1800 and 6.4 billion in 2000).

Migration redistributes the genetic, physiological, morphological, and sociocultural differences found in human populations. Thus, it is likely that rural-to-urban migration would have some effect on the growth and development of migrants, and the recipient population. Livi (1896) and Ammon (1899) (cited in Boas, 1912; 1922) were the first to publish studies of the growth of urban migrants. Livi found that the children of urban migrants in Italy were taller than rural sedentes (the non-migrating rural population). He believed the reason for this was heterosis, the marriage of urban migrants from different rural regions leading to 'genetic vitality' in their offspring. Ammon also found the children of migrants to be taller than rural sedentes, but he argued for the action of natural selection to explain this. Though it is not clear what type of selection pressure was supposed to be involved, perhaps he meant that, in the rigors of the urban environment, only the 'fittest' (the tallest?) would survive.

These views that a genetic mechanism, heterosis or natural selection, was at work stem directly from the erroneous belief that human types are genetically fixed, that types will not change when exposed to different environments. This belief was shattered by the publication of Boas' study of the 'Changes in the bodily form of decendants of immigrants' (1912). Boas found that the children of immigrants to the United States were taller and differently shaped than their parents and the non-migrating populations from which their parents came. He stated that neither natural selection nor heterosis could adequately account for these changes. Rather, modifications in the process of growth and development as a response to environmental change were responsible. Boas was vigorously attacked for this position. In a series of papers he whittled away at his attackers (see Boas, 1940). His evidence against the fixity of types was that: (1) the physical differences between parents and children appeared early in the life of the child and persisted until adulthood, (2) the longer the childhood exposure to the American urban environment the greater the physical difference, (3) children from large families were shorter than children from smaller families of the same 'racial type,' and (4) differences between parents and children were greater when botn were foreign born, meaning that only the children would have been exposed to the new environment during the developmental years, than when the child was American born, meaning that the parents may have spent some of their growth years in the United States.

The now classic studies of Shapiro (1939) on the growth of Japanese chidren in Japan and Hawaii, and of Goldstein (1943) and Lasker (1952) on the growth of Mexicans in Mexico and the United States confirmed the nature of human developmental plasticity. Today, human developmental plasticity is taken for granted, but it was the migration research of Boas, and those following his lead, that established the validity of this phenomenon. Shapiro's Japanese migrant study compared the growth of Hawaiian-born Japanese of Japanese immigrant parentage, Japan-born Japanese who migrated to Hawaii, and Japanese sedentes living in the same villages from which the migrants originated. The sedentes were mostly farmers and laborers in rural villages. The exact location of the migrants was not given, but 65 per cent of the recent immigrants were employed in sedentary occupations (store clerks), and 76 per cent of the Hawaiian-born were either students or sedentary workers. The immigrants appear to have been developing an urban lifestyle. The sedentes and the recent immigrants differed in a few anthropometric measurements, some increasing and some decreasing with migration. The largest differences were between the immigrants and the Hawaiian-born. The latter were taller and more linear in body build than their parents or the sedentes. Shapiro argued that, with migration, there were improvements in diet, health care, and socioeconomic status and that these conditions, associated with an urban lifestyle, were responsible for the growth changes.

A well-conceived study of rural-to-urban migration and growth was carried out in Poland. The results were published in Polish by Panek & Piasecki (1971) and were summarized in English by Tanner & Eveleth (1976). A new industrial town was created on the outskirts of Cracow in 1949. In 1965 the population reached 1599 persons per square kilometer, similar to that of many other European cities. Most of the population growth was due to migration from rural villages. The children and youth measured for the study had been born in the new city or had lived there at least 10 years. The children living in the new city were, on average, five to six centimeters taller and one to two kilograms heavier than the rural sedentes. The urban children matured earlier, both for tooth eruption and age of menarche. Thus, even in the post-World War II period, the children of urban migrants experienced significant improvements in growth.

Reviews of the literature by Tanner & Eveleth (1976) and Malina (1979) show that healthy, well-nourished children living in the cities of the less developed countries are usually taller and mature earlier, as measured by rates of skeletal development and age at menarche, than their rural age peers. Bogin & MacVean (1981b) found that mildly malnourished urban school children, from low socioeconomic status families, living in Guatemala were taller than rural children of the same age. However, migrants to urban slums in less developed countries do not usually experience the benefits of the urban environment. In Asia, Africa, and Latin America the slums are often on the outskirts of the cities. As squatter settlements they have no official access to city services and facilities. Not surprisingly, the growth of migrant children living in these slums is not significantly different from that of children living in the impoverished rural areas. Indeed, Malina *et al.* (1981) showed that children living in the urban slums of Oaxaca, Mexico were significantly shorter and lighter than nearby rural children from a reasonably prosperous town. Johnston *et al.* (1985) showed that residents of an impoverished community on the outskirts of Guatemala City were no taller or heavier than rural populations in Guatemala.

Though the environment for growth in the city, versus the countryside, may determine the size of children, it has also been proposed that a biological selection of individuals takes place for migration and for marriage. By biological selection it is meant that migrants may not be a random sample of the rural population. Migrants may be genetically taller, mature more rapidly, or differ in other physical aspects from rural sedentes. Positive assortative mating between such people may further differentiate migrants from sedentes.

Several studies provide tentative support for biological selection. Steegmann's (1985) historical study of 18th century British military recruits found that conscripts who migrated from their county of birth were significantly taller (169.1 cm) than recruits living in the county of their birth (167.6 cm). Recall though, that he also found that urban-born recruits were shorter than rural-born men. Steegmann pointed out that 18th century Britain was a developing country and that urban areas were characterized by food shortages and unhygienic living conditions. Thus

climate, or migration. However, socioeconomic status has an equally important impact on growth. For instance, Bielicki *et al.* (1981) studied the stature of 13 000 Polish military recruits, born in 1957 and measured in 1976, in relation to three environmental factors: occupation and education of the father (SES), population size of the conscript's home city, town, or village (urbanization), and number of siblings in the conscript's family (family size). Mean stature decreased with lower SES, less urbanization, and larger family size. Though each factor was shown to have an independent and significant effect on stature, the strongest effect was for SES, followed by family size, and then urbanization.

The human populations of southern Mexico and Guatemala may be studied to assess the influence of several variables on child growth. These populations include people of Maya Indian and Ladino (mixed Spanish and Indian) ethnicity, people engaged in a continuum of occupations from subsistence agriculture (low SES) to professional and technical jobs (high SES), rural and urban residents, and citizens of two different nations. Using a multifactorial approach, the relative influence of each of these variables on the growth status of Mexican and Guatemalan children can be computed. The following analysis uses data from published sources for six groups of children. Three of the groups, rural Zapotec-speaking Indians, rural Ladinos, and urban Ladinos from the state of Oaxaca, Mexico, were studied by Malina *et al.* (1981). The Zapotecs are subsistence farmers, with some income from crafts and wage labor. The rural Ladinos are cash-crop agriculturalists, and differ from their Zapotec neighbors in their more 'westernized' life style, clothing, agricultural techniques, and formal education. The urban Ladinos live in a *colonia,* a very low socioeconomic status neighborhood, and parents of the children work as unskilled or semi-skilled laborers. The three Guatemalan groups, urban Ladino children of very high SES, urban Ladinos of low SES, and rural Cakchiquel-speaking Indian children of low SES, were studied by Bogin & MacVean (1978; 1981a; 1984). The high SES urban Ladinos are from some of the wealthiest families in Guatemala and are afforded all of the health, nutritional, and social benefits of their privileged economic status. The low SES urban Ladinos are children of low-paid semi-skilled laborers. The families of the rural Cakchiquels follow traditional Indian cultural practices in terms of language, clothing, house construction, and many social behaviors. However, only four per cent are agriculturalists, the traditional occupation; the remainder are small-business persons engaged in the production and sale of crafts and clothing, or are vendors of agricultural produce.

Means and standard deviations for the height, weight, arm circumference, triceps skinfold, and arm muscle circumference of each group are given in Table 5.1. The data represent average values for children aged 7.00 to 13.99 years old. Children of higher SES, from urban areas, and of Ladino ethnicity are generally larger, in most dimensions, than children of lower SES, from rural areas, or of Indian ethnicity. The most significant difference between samples is for the triceps skinfold measure, an indicator of energy reserves (i.e., adipose tissue) and, hence, nutritional adequacy. Diet surveys conducted in Mexico and Guatemala find that intakes of protein and calories are below national standards for rural and Indian populations (Bogin & MacVean, 1984). The skinfold data reported in Table 5.1 confirm the poor nutritional status of the low SES groups.

A canonical correlation analysis was used to calculate the rank order of importance of each of the independent variables (parental occupation, rural–urban residence, ethnicity, and nationality) for growth. This analysis reduces groups of related variables to statistical 'factors.' In this case, the five anthropometric measurements (stature, weight, etc.) were reduced to one factor, which may be called 'growth status.' The correlation between the 'growth status' factor and each of the independent variables was calculated, and from the values of the correlation a canonical score for each independent variable was derived. The absolute value of the canonical score is a measure of the relative influence of parental occupation, rural–urban residence, ethnicity, or nationality on 'growth status.' The canonical scores ranked in the following order: rural–urban residency = 1.68, parental occupation (SES) = 1.43, Mexican–Guatemalan nationality = 0.98, and Indian–Ladino ethnicity = 0.90. Clearly, the importance of factors associated with the general social and economic environment transcends the influence of ethnicity or nationality, and their possible genetic or cultural concomitants, as determinants of the growth of these samples of children.

In developing nations, such as Mexico and Guatemala, there are great differences in the quality of the environment for growth from rural to urban areas, and from low to high SES. In the developed nations of North America, Western Europe, and Japan, public and private systems of health care delivery, food distribution, and social services may significantly reduce environmental differences between rural and urban regions and SES classes. Even so, the power of the socioeconomic effect on growth is still in evidence. For instance, variation in height due to region of origin and father's occupation (used as a measure of SES) was analyzed by Mascie-Taylor & Boldsen (1985) in a national sample of 33 000 11-year old British children and their parents. Children and adults from Scotland, Wales, and northern England were found to be shorter than subjects living in central and southern regions of England. However, statistical

Table 5.1. *Means ($\bar{x}$) and standard deviations (sd) for growth status variables of children from Oaxaca, Mexico and Guatemala*

		Guatemalan samples						Oaxaca samples					
		Urban Guatemalan High SES		Urban Guatemalan Low SES		Cakchiquel Indian		Rural Zapotec		Rural Ladino		Urban Colonia	
Growth variable		Boys	Girls	Boys	Girls	Boys	Girls	Boys	Girls	Boys	Girls	Boys	Girls
Height (cm)	$\bar{x}$	138.40	138.37	130.53	131.00	126.76	125.50	125.59	127.13	129.31	129.14	127.63	127.47
	sd	(6.56)	(7.83)	(6.71)	(6.64)	(5.93)	(6.10)	(5.66)	(6.27)	(5.61)	(5.76)	(6.04)	(5.79)
Weight (kg)	$\bar{x}$	34.31	35.77	29.36	29.83	27.37	27.23	25.43	27.00	27.33	28.00	27.10	27.27
	sd	(6.56)	(7.75)	(5.39)	(5.33)	(4.07)	(4.33)	(3.26)	(3.69)	(3.59)	(3.86)	(4.23)	(4.13)
Arm circumference (cm)	$\bar{x}$	22.20	22.07	19.56	19.89	18.59	18.47	17.43	18.26	18.54	19.29	18.80	19.04
	sd	(2.59)	(2.07)	(1.80)	(1.81)	(1.53)	(1.40)	(1.36)	(1.43)	(1.43)	(1.41)	(1.66)	(1.60)
Triceps skinfold (mm)	$\bar{x}$	11.23	13.37	8.47	10.51	6.34	7.67	6.16	8.76	5.46	8.14	6.27	8.13
	sd	(5.03)	(4.61)	(2.87)	(3.44)	(2.34)	(2.66)	(1.41)	(2.01)	(1.64)	(2.33)	(2.24)	(2.63)
Arm muscle circumference (cm)	$\bar{x}$	18.67	17.86	16.86	16.54	16.60	16.04	15.49	15.50	16.84	16.73	16.81	16.50
	sd	(1.47)	(1.35)	(1.38)	(1.35)	(1.26)	(1.16)	(1.22)	(1.19)	(1.17)	(1.14)	(1.31)	(1.14)

Note: Ages 7–13 were combined to compute the means and standard deviations.

analysis of the data showed that the regional effect could be explained, almost entirely, by variation in the occupation of the father. In central and southern England, fathers had significantly higher occupational status than in the other regions. Within Great Britain, geographic variations in climate, diet, and genetic growth potential are subordinate to socioeconomic status as determinants of growth in height.

A number of studies have established that in the United States and Western Europe, children and adults of lower SES are, on average, shorter, and have less muscle mass and skeletal mass, but more fat mass, than individuals of higher SES (Garn & Clark, 1975; Fulwood *et al.*, 1981; Clegg, 1982; Malina *et al.*, 1983). Garn *et al.* (1977) found that poorer, leaner girls in the United States grow up to be poorer, fatter women who have babies of lower birth weight than women of higher SES. Because birth weight is a powerful determinant of subsequent growth, the authors wondered whether the SES differences found in later life might not be due to differences in prenatal growth and birth weight. The question was addressed by Garn *et al.* (1984) in a study of more than 3000 live-born singletons from a national sample of births in the United States. Using a composite score of parental education, occupation, and income to measure SES, the sample was composed of two SES groups: low, with scores from one (minimum) to three, and high, with scores from eight to 10 (maximum). After matching, or correcting, for maternal prepregnancy weight and birth weight, it was found that low SES children were significantly shorter, weighed less, and had smaller head circumference than the high SES children. The differences were present at birth and increased each year after birth up to seven years of age. The postnatal SES effect was also found to influence mental development. Garn *et al.* selected a sample of low and high SES infants, all eight months old, with scores of 80 or higher on the Bayley scales of infant development. Scores above 80 indicate satisfactory motor and mental development. General intelligence (IQ) tests were given to the children at four and seven years of age. At these ages, about 25 per cent of the low SES children had low IQ scores, while only about seven per cent of the high SES children had low scores. These findings showed that the SES differences in physical and cognitive growth were not due to prenatal and perinatal influences alone, rather the socioeconomic effects were cumulative through seven years of age. Of course, this finding is not unexpected since it is known, as shown in the previous chapter, that the infant and childhood periods of growth are highly sensitive to many environmental determinants of growth.

Even within a seemingly homogeneous population, subtle differences in SES have a significant influence on growth. Johnston *et al.* (1980)

analyzed the longitudinal growth records of 276 rural Mexican children. All the children came from families practicing traditional subsistence agriculture, with minimal opportunities for wage labor. The sample of children was divided into two groups, one with evidence of chronic undernutrition and growth failure, and another with satisfactory nutrition and growth status. Out of 38 variables measured for their impact on growth and nutrition, three were significant: socioeconomic status, father's linear size, and mother's linear size. Malnourished and poorly growing children came from poorer families, with less well educated parents, and had parents with smaller linear dimensions (height, biacromial breadth, etc.) than better nourished children. Since small body size is likely to be the result of poor living conditions (see previous sections of this chapter), the effect of parental linear dimensions may itself be due to the low SES environment of the parents when they were children. Garn *et al.* (1984) called this transgenerational influence of low SES the effect of 'recycling of poverty.'

Similar socioeconomic variation in growth was found by Malina *et al.* (1985) for 293 children from a rural Mexican village, measured cross-sectionally. The subjects were Zapotec-speaking Indians, six to 13 years old, living in families that practiced subsistence agriculture. Previous work in the community showed that nutritional status was marginal, average dietary intakes being below the recommended levels for Mexico, and that infant mortality rates were about three times the national average. Thus, the village as a whole suffered from poor environmental conditions for health and growth. An SES index for families in the village was constructed from landholdings, household goods, and parental occupation. About 42 per cent of the families were landless, about 40 per cent did not own a radio or other household appliance, and about 89 per cent of the parents were engaged in full-time farming. The SES index was found to reflect child health and mortality. Mortality prior to age 15 years was about 18 per cent of the 141 live births in the higher SES group, and in the lower SES group it was about 22 per cent of the 152 live births. Boys from better-off families, that is, families with some land, household appliances, or non-agricultural occupations, were significantly heavier and had thicker triceps skinfolds than boys from poorer families. No differences between SES groups were found for girls. These findings indicate that boys were able to benefit from improved living conditions in this rural Indian community, even though the better-off families were still living at or below poverty level. It was not known why the girls did not show an SES effect on growth. Perhaps child rearing practices, such as feeding schedules and health care, differed according to the sex of the child.

A girl's age at menarche is known to be a sensitive indicator of environmental conditions of growth, and girls from lower SES back-grounds usually have a later mean age at menarche than girls from higher SES backgrounds (Johnston, 1974). The subtleties of the SES–menarche relationship were reviewed by Bielicki & Welon (1982) for Polish popula-tions. When parental education, family size, and nutritional status were controlled statistically, menarche was found to occur later in girls whose fathers lived in small towns rather than in big cities, and in girls whose fathers had lived in a city for less than 15 years, rather than for 15 to 20 years. Daughters of full-time farmers reached menarche later than daughters of part-time farmers living in the same villages. Daughters of men with an elementary school education had later menarche than daughters of men with elementary school and two years of vocational school educations. These findings show, once again, how powerful the SES effect may be, even when living conditions vary only slightly between groups of people.

Two other studies of maturation and SES show the power of this relationship. Low *et al.* (1982) found, in the period 1961–3, significant differences in the sexual maturation (breast and pubic hair development, and age at menarche) of Hong Kong girls of high and low SES. A follow-up study during 1977–9 found no SES differences in sexual matura-tion. The change was entirely due to the more rapid development of the low SES girls during the more recent period. The authors pointed out that social and political changes, resulting in higher wages for unskilled laborers and the availability of low-cost, decent public housing, were the major environmental improvements between the earlier and later studies. Thus, socioeconomic improvements that are, at best, indirectly aimed at children can have a direct influence on human development.

The other study, by Hoshi & Kouchi (1981), is an assessment of changes in the age at menarche in Japan from 1884 to 1980. Mean menarchial age was about 15 years during the period 1884–1920. By 1940 the age had dropped to about 14.2 years, a decrease similar in magnitude to that found for European and North American girls. World War II reversed this trend, so that by 1952 mean age at menarche returned to 15 years. Although the war had ended by this time, these girls were conceived and born just prior to the war years, and passed through their environmentally sensitive periods of infant and childhood growth during the war. There was a rapid decline in mean age at menarche after 1952, so that in 1959 the age was 13.3 years. Girls of that mean age were born just after the war ended, and grew up during the reconstruction period following the war. In recent years the decreasing trend has slowed, and in 1980 the average age of menarche was 12.4 years, which is very close to the average age for well-off girls from other populations. As previously described, the post-war period in Japan was one of tremendous social, economic, political, and dietary change. The rapid decline in the age at menarche immediately after the war, when these changes were most drastic, and the slowing of this decline more recently, shows how closely linked are the socioeconomic development of a nation and the physical development of its children.

What are the direct causes of the relationship between socioeconomic status and human growth? Malina (1979) argued that better health care, which reduces childhood mortality and morbidity, also is the most important cause of increased growth associated with higher SES. Bielicki & Welon (1982) listed four primary factors: higher SES allows for better nutrition, better health care, reduced physical labor for children, and greater growth-promoting psychological stimulation from parents, schools, and peers. the relation of the first three to growth has been adequately described in preceding sections of this discussion. Bielicki & Welon did not explain how psychological stimulation could increase growth. Descriptive observations, combined with experimental studies for support, are needed to interpret the impact of psychological factors on physical growth.

Matsumoto (1982) found that urbanization and income not needed for food purchases were associated with faster rates of growth. He analyzed changes in the rate of growth in height of Japanese children born in the years 1888 to 1962. The age at peak height velocity (PHV) during the adolescent growth spurt was used to assess the impact of environmental factors on growth rate. As expected, World War II had a delaying effect on growth rate; on average, boys born in 1894 reached PHV at about 14 years, boys born in 1936 reached PHV at about 14.9 years. After the war, the average age at PHV lowered rapidly and reached about 13 years for boys born in 1950–5 and 12.8 years for boys born after 1960. Similar changes occurred for girls, who generally reach PHV about two years earlier than boys.

Matsumoto evaluated the influence of six independent variables on rate of growth: diet (calories from starchy foods and calories from fat), family size, industrialization, national income level, urban population rate (number of people living in cities), and Engel's coefficient (ratio of food expenses to total living expenses). Using the statistical technique of lagged correlations, the time difference between year of birth, the change in the independent variables, and the trend in the age at PHV were calculated. It was found that diet, family size, industrialization, and national income level had maximum effects on age at PHV after a 16 to 24 year lag from year of birth. Since the average age at PHV occurred before age 16 in all birth cohorts, it was unlikely that any of these variables was a primary cause of the changes in rate of growth. In contrast, the lag between year of birth and urban population rate, Engel's coefficient, and age at PHV was eight to 10 years. Thus, the increase in the number of people living in cities and the decrease in the amount of money spent on food relative to other living expenses preceded the decrease in age at PHV and may have caused it.

Modern cities have a wider variety of sanitation, health, educational, and social services, and generally offer a wider variety of foods, at lower prices, than towns and villages. Urban children, whose parents are engaged in wage labor, work less at energy intensive labor than rural children, whose parents are engaged in agricultural labor (Bogin, 1987). Living in a city and having income beyond that spent on food allow people to avail themselves of these urban services and benefits. The calories saved by better health and reduced child labor, plus the availability of more food and more calorie dense foods (such as those processed with added sugar and fats) are invested into bigger, faster growing children. It is in this way that small increments in education and occupation, that translate into higher income and greater knowledge and use of city services, are associated with growth.

Where urban lifestyles have spread to the countryside, such as in much of the United States and England, urban–rural differences in growth have disappeared. However, within the urban or rural areas, children of the very poor do not benefit from the opportunities of urbanization. The studies described above by Rona & Chinn (1986) in England and Malina *et al.* (1981) in Oaxaca, Mexico showed that children from the lowest SES segments of the city were smaller than national averages or rural living children. A number of studies of rural populations, summarized by Tobias (1985), show that there has been an absence of growth change or a trend for negative growth in low SES groups around the world. Absent or negative trends were cited by Tobias for East Africa, southern Africa, Korea, the Middle East, Turkey, India, and Latin America. The causes of such trends were, in each case, linked to poor environmental conditions. For example, a deterioration of the economic and social structure of the Zulu and Tonga cultures in southern Africa since the colonial period (early 1900s) was accompanied by a decrease in adult stature. Working with skeletal populations, Tobias & Netscher (1976) showed a decrease in femur length for South African blacks from 1925 to 1973, a period of social and economic decay for these people.

The poor living conditions, including high rates of undernutrition, infectious disease, childhood mortality, and heavy workloads, of Amerindian populations of Mexico and Guatemala have not changed appreciably in the last 100 years. Consequently, the small average stature of children and adults, and the slow rate of maturation of children, have remained constant during that time (Malina *et al.,* 1980; McCullough, 1982; Bogin & MacVean, 1984). This has occurred despite national and international efforts to promote the economic, social, and political development of Mexico and Guatemala, efforts designed to improve living conditions in these countries.

The Indian populations of Mexico and Guatemala, like the blacks of South Africa, have been systematically denied a share in the development and resources of their countries. The poor growth of the children and adults of these disenfranchized peoples is an unassailable testament of the discrimination practiced against them. Virtually every nation is guilty of racial discrimination. Even in developed nations, such as the United States and England, ethnic minorities are overrepresented, as a proportion of the total population, in the lowest SES levels of society and in the poorest growing groups of children.

In recent years, only two nations, Norway and Sweden, seem to have eliminated differences in growth between children from higher and lower SES levels. Brundtland & Walløe (1973) and Brundtland *et al.* (1980) found that for children born after 1955 there were no differences in average age at menarche of Norwegian girls or in average height of boys and girls from different SES backgrounds. Children from lower SES families weighed more than children from higher SES families, a common finding for developed industrial nations. Lindgren (1976) found no differences in average height and weight, or in average age at peak height velocity, peak weight velocity, and menarche in a sample of 740 Swedish urban school children. Socioeconomic differences, as measured by education, occupation, and income, do exist in these Scandinavian nations, but federal systems of guaranteed health care and social support services provide people of all SES levels with an equal share in the environmental opportunities for growth and development.

There is another side to this issue of socioeconomic effects and growth, which is that within all social classes taller individuals tend to move up in SES and shorter individuals tend to move down or remain stable in SES. Stature related social mobility even occurs within families. Some aspects of this phenomenon were discussed above in the section 'Migration and urbanization.' It may be recalled that, on a population basis, migrants tend to be taller and of higher SES than sedentes. On an individual basis, Scott *et al.* (1956) found that taller women from Aberdeen, Scotland worked at more skilled jobs than shorter women. Furthermore, when taller women married, their husbands worked at more skilled jobs than the husbands of shorter women. These findings were confirmed by Schreider (1964b) in a study of British women showing that, regardless of the occupation of the father, taller women tended to marry men with more highly skilled jobs and shorter women tended to marry men performing semi-skilled or unskilled labor.

Taller people achieve higher educational status than shorter people, as was shown by Parnell (1954) for university students in England and other European countries and by Kimura (1984) for Japanese university students. Susanne (1980), working in Belgium, and Bielicki & Charzewski (1983), working in Poland, found that within families the taller siblings were better educated than shorter siblings. In Belgium, there was a higher correlation between a young man's stature and his level of education than between the same individual's level of education and his father's occupation. In the Polish study, the average difference in stature between better and less well educated brothers ($n = 116$ pairs) was 1.26 cm, a statistically significant difference. Better education, of course, is likely to lead to more skilled and higher paying employment, and, consequently, higher SES.

These studies show that the tall tend to rise in SES and the short tend to sink. Stature, by itself, does not determine occupation, education, or socioeconomic attainment, but there is a strong social bias in favor of the tall which may help facilitate their SES climb. Keyes (1979) found in a questionnaire survey that this bias operates in the United States, and probably other Western nations. Men under 175 cm in stature, the average height for men in the United States, 'invariably wished they were taller.' Keyes reviewed other research and found that job status and economic rewards could account for this desire. In two studies, conducted three years apart, of graduates from the University of Pittsburgh, there was a 12.4 per cent difference in initial starting salary favoring men 188 cm tall versus men 180 cm tall. The salary difference favoring *cum laude* graduates was only four per cent. Additional research on 5000 US Army recruits measured in 1943 found that, in 1968, those over 183 cm earned eight per cent more than those below 168 cm, even after the influences of IQ, educational level, marital status, and occupation were statistically controlled. In another study, 140 personnel officers of companies involved in retail sales were asked which of two equally qualified job applicants would they choose – one who is 185 cm tall or one who is 165 cm. The taller applicant was preferred by 72 per cent of the personnel officers, 27 per cent had no preference, and one respondent chose the shorter applicant.

A similar bias operates among the Mehinaku Indians of Brazil, a traditional culture based on subsistence farming and fishing. Gregor (1979) found that taller Mehinaku men had more wives and lovers, more wealth, and more social prestige than shorter men, and that the latter were often the objects of social and sexual ridicule. It is interesting to note that the range of stature for Mehinaku men observed by Gregor was 151.8 cm to 175.9 cm, meaning that the tallest men would be considered about average in height if United States values of tallness and shortness were applied. Gregor reviewed ethnographic data from several other cultures and found that only the Crow of North America and the Mbuti pygmies of Central Africa specifically mention tallness as a disadvantage.

Marriage brokers in Shanghai, even today, report that greater stature is the first consideration of women seeking a husband. Only high status of his occupation or a home of his own can offset being short in the market for marriages (Lasker, personal communication).

It is possible to speculate about the historical and psychological reasons why the 'bigger is better' bias exists in many societies. The concrete result of this bias, just as for the discrimination stemming from racial prejudice, is that individuals or groups of the accepted 'type' are more likely to receive better care, in the widest sense, than individuals or groups of the undesired type. A positive feedback relationship between growth and socioeconomic status results from the social bias; better environmental conditions lead to larger size, taller individuals tend to rise in SES, and higher SES leads to better environmental conditions. An opposing cycle exists for those from lower SES, or shorter individuals from any social class. The result is that differences in physical size between individuals are both a consequence and a cause of socioeconomic effects on growth.

6 Genetic and endocrine regulation of human growth

Children tend to resemble their parents in stature, body proportions, and rate of development. It may be assumed that, barring the action of obvious environmental determinants of growth, these resemblances reflect the influence of genes that parents contribute to their biological offspring. Genes do not directly cause growth and development. Rather, the expression of a genetically inherited pattern of growth is mediated by several biological systems, which must operate within an environment appropriate for growth. An example is the endocrine system. Genes may regulate the production and release of hormones, such as growth hormone or testosterone from the endocrine glands, that stimulate the growth of cells and the development of tissues toward their mature state. The various growth regulating hormones may be released sequentially, simultaneously, or not at all. It is at this level of endocrine interaction that the stimuli that cause growth may actually be produced. The endocrine system also responds to the influence of many environmental factors that affect human development, and may, therefore, serve as a mechanism that unifies the interaction of genes and the environment to shape the pattern of growth of every human being. This chapter describes the nature of the genetics of growth, the endocrine system, and the effect of the interaction of genes, the environment, and hormones on human development.

Genetics of human development
The influence of heredity on the growth and development of individual children was addressed by Bock (1986). He analyzed data from children, raised under favorable environmental conditions, participating in one of the major longitudinal studies of human growth conducted in the United States. Starting in the 1930s, the growth of healthy, well-nourished children, living in small urban communities and rural areas of southwestern Ohio, has been measured by researchers working at the Fels Research Institute. Participants in the study are measured longitudinally, ideally once a year or more often, from birth to maturity for height, weight, and a variety of other physical and psychological characteristics. As of 1983, 214 boys and 234 girls had been measured to maturity. The

160

families of the Fels sample are largely of middle socioeconomic class and of European cultural background. Thus, the sample is not representative of all children of the state of Ohio or of the United States, however the Fels study provides a wealth of data about the growth and development of normal children.

Bock analyzed the data for several individuals from this study, cases representing the extreme variants in the normal range of growth and development. His purpose was to describe variation in the inheritance of patterns of growth. In Figure 6.1, the height and velocity of growth in height of the tallest girl and the earliest maturing girl (defined by the rate of skeletal maturation) are compared. Both girls were tall for age throughout infancy and childhood, being above the 95th percentile for height for all girls in the Fels sample. The early maturing girl entered the adolescent growth period in her seventh year, reached peak height velocity at 9.4 years, and stopped growing by age 13 years, when she reached the 75th percentile of height. The skeletal maturation of this girl, as estimated from hand-wrist radiographs, was advanced over her chronological age by about three years. Thus, her tallness during child-hood was due, in large part, to her advanced maturation. The tall girl entered adolescence at about 10.0 years, reached peak height velocity at 12.2 years, and ceased growing at about 18 years at a height of 185.1 cm, which was at the 99th percentile. The parents of this girl were also tall, compared with other adults of like sex and age, which led Bock to state

Figure 6.1. Distance and velocity curves for the growth in height of the tallest girl (solid lines) and the earliest maturing girl (dashed lines) participating in the Fels Research Institute Study of growth (after Bock, 1986).

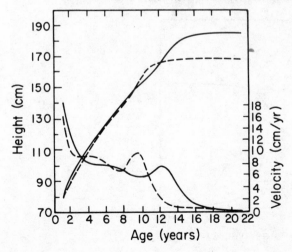

that the growth of this girl is a 'typical example of intrinsic tall stature of familial origin.' Bock believes that the girl's tallness is a genetically determined characteristic, inherited from her parents.

Presented in Figure 6.2 are growth curves for the shortest girl and the latest maturing girl (again, defined by rate of skeletal maturation), measured so far in the Fels study. During childhood both girls were of similar stature, being at about the fifth percentile for the Fels sample. The short girl entered adolescence just a little later than the earliest maturing girl (shown in Figure 6.1), stopped growing in her fourteenth year, and reached 149.0 cm, which is below the fifth percentile. This girl was advanced in skeletal maturation from age eight to age 16, and Bock explained that her short stature was a family characteristic, inherited from her parents, of limited growth in size, combined with a developmental pattern for early maturation. The late maturing girl entered adolescence about two years later than the short girl and, with this extra time for prepubertal growth, reached 160.1 cm which was just under the 50th percentile. Her skeletal maturation was delayed in comparison with her chronological age. The delay fluctuated between one-half year and two years throughout her childhood and adolescence. This slow rate of development allowed for a prolonged growth period and attainment of average stature.

Bock described these four cases of growth as 'unusual,' meaning that they represent the limits of the range of normal variation in amounts and

Figure 6.2. Distance and velocity curves for the growth in height of the shortest girl (solid lines) and the latest maturing girl (dashed lines) participating in the Fels Research Institute Study of growth (after Bock, 1986).

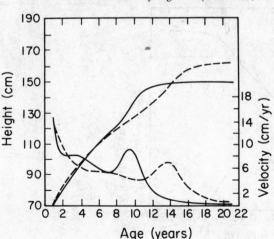

rates of growth for the sample of the Fels Research Institute. The four girls were raised under favorable environmental conditions, and showed no evidence of acute or chronic diseases that influence growth and development. Given this, and measurements of height of the parents of each girl, Bock's analysis indicates a major genetic component in the determination of size and rate of growth in these girls.

Twin studies as an approach to the genetics of growth

Twin studies offer a more direct methodology for delineating the influence of heredity on growth. In Tables 6.1 and 6.2, data are presented that compare the correlations in height and weight between monozygotic (MZ) and dizygotic (DZ) twin pairs at ages from birth to eight years. The data are from the Louisville, Kentucky Twin Study and comprise 'over 900 twins who have been recruited and measured since 1962' (Wilson, 1979).

If amounts and rates of growth are totally controlled by the genotype, then correlation coefficients for MZ twins, who are genetically identical, should be equal to 1.00, a perfect positive correlation, at all ages. The correlation coefficient for DZ twins, who share an average of 50 per cent of those genes that are free to vary, should be equal to 0.50 at all ages. Of course, this assumes that parents are randomly selected from the population of potential mates. This is usually not the case, for instance, positive assortative mating for height occurs in many human populations. Even so, the data in Tables 6.1 and 6.2 show that by one year of age, and thereafter, the correlation coefficients predicted from the genetic model are very nearly found for this relatively large sample of twins.

Table 6.1. *Correlation coefficients for height between monozygotic (MZ) and dizygotic (DZ) twin pairs from birth to age eight*

Age	Total n	MZ	DZ Same sex	DZ Different sex
Birth	629	0.62	0.79	0.67
3 months	764	0.78	0.72	0.65
6 months	819	0.80	0.67	0.62
12 months	827	0.86	0.66	0.58
24 months	687	0.89	0.54	0.61
3 years	699	0.93	0.56	0.60
5 years	606	0.94	0.51	0.68
8 years	444	0.94	0.49	0.65

From Wilson (1979).

Table 6.2. *Correlation coefficients for weight between monozygotic (MZ) and dizygotic (DZ) twin pairs from birth to eight years*

Age	Total n	MZ	DZ Same sex	Different sex
Birth	992	0.63	0.68	0.64
3 months	766	0.74	0.66	0.40
6 months	819	0.81	0.63	0.39
12 months	828	0.88	0.55	0.37
24 months	779	0.88	0.53	0.50
3 years	713	0.88	0.52	0.54
5 years	606	0.85	0.48	0.62
8 years	444	0.88	0.49	0.46

From Wilson (1979).

The coefficients from birth to six months of age, however, do not correspond with genetic expectations. At birth MZ twins are less concordant in size compared with DZ twins. One reason for this may be that MZ twins often share a monochorionic placenta during the prenatal period. Falkner (1978) reviewed data from seven studies of twin placentation, and found that about 70 per cent of MZ twins have monochorionic placentae. Vascular anastomoses (arterial and venous connections) between the parts of the placenta supplying blood to each twin occur in monochorionic placentae. This results in a transfusion of blood, and the oxygen and nutrients carried by the blood, between the twins. The transfusion of blood is usually not equal, which means that the twins do not receive an equal maternal blood supply, possibly resulting in undernutrition and hypoxia (low oxygen availability) for the disadvantaged twin. In the condition of dichorionic placentation, each twin receives a separate supply of maternal blood and nutrients. Falkner (1966) found that in a sample of 92 MZ twins, the within-pair difference in birth weight averaged 326.0 grams in monochorionic twins and 227.8 grams in dichorionic twins. The significantly larger difference in monochorionic MZ twins suggests that the fetus exposed to less of the maternal blood supply may have to adjust its intra-uterine growth rate to adapt to placental insufficiency. This would lower correlation coefficients in length and weight at birth. The values of the correlation coefficients at birth for DZ twins, displayed in Tables 6.1 and 6.2, are higher than expected from a simple genetic model of growth. DZ twins are more likely to have separate placental connections with the mother's vascular system. In this placental condition, it is more likely that each twin receives a relatively equal share of oxygen, nutrients, and other substances from the mother. This may explain, in part, the relatively high concordance in size of the DZ twins at birth. Robson (1978) reviewed birth weight statistics for several populations of people living in the developed nations and found that, for singleton or DZ twin pregnancies, up to 66 per cent of the variance in birth weight is due to maternal environment. That environment includes the quality of the mother's diet, smoking habits, alcohol usage, and other variables of this type that influence the quantity and quality of the maternal blood supply to the fetus. Robson estimated that genetic factors account for about 34 per cent of the variance in birth weight, and of this only ten per cent of the variance is due to fetal genotype. Thus, the higher than expected concordance in the birth weight of DZ twins is likely to be due to the shared maternal environment, which sets some common limits to the growth of both twins. After parturition, growth is determined by the interaction of the environment with the unique genotype of each DZ twin. This probably explains why the correlation coefficients for DZ twins in the Louisville, Kentucky Twin Study become less concordant as the twins get older.

In some instances, prenatal influences on growth have effects that last for many years after birth and may obscure the contribution of genes to the determination of size. Wilson (1979) examined a sub-sample of MZ twins from the Louisville study. This sub-sample consisted of 10 MZ twin pairs with the largest differences in birth weight. The lighter twin averaged 57 per cent of the birth weight of the heavier twin and the average absolute difference between twins in birth weight equalled 1,064 grams. The smaller twin was usually of low birth weight and required special postnatal care. Wilson found that by six years of age the relatively large initial disparities were progressively reduced but not eliminated. The mean difference at six years was 2.19 kg for weight and 1.85 cm for height, indicating that the lighter twin at birth was still ten per cent lighter and two per cent shorter, on average. For these 10 pairs of MZ twins the within-pair correlation for height was only 0.72 compared with the correlation of 0.93, or higher, for all MZ twins in the Louisville study (Table 6.1).

Similar long-term differences in growth between MZ twins of markedly different birth weight have been reported by Falkner (1966; 1978). In one case, full-term MZ twin boys had birth weights of 1,460 grams and 2,806 grams and birth lengths of 43 cm and 50 cm respectively. The smaller twin was considered small for gestational age, and required special neonatal care. Inspection of the placenta showed that it was monochorionic, with only about 40 per cent of the placenta supplying maternal blood to the smaller twin. During the first year of life, the growth rate of the smaller twin exceeded that of the larger twin, but the catch-up in growth was incomplete. Differences in height, weight, and other physical measurements persisted at all ages through 16 years of age, e.g., heights and weights at age 16 were 161.9 cm and 50.6 kg for the smaller twin and 167.3 cm and 58.5 kg for the larger twin.

The implication of these observations is that twin studies may provide clear indications of the genetic control of growth, but only when the environment for growth is favorable or at least does not inhibit the growth of one or both twins.

Correlations in growth between biological relatives (non-twins)

Studies of familial correlations in growth may help clarify the role of genes and the environment. Familial correlations for serial measurements of stature were analyzed by Byard *et al.* (1983), using data from the Fels Research Institute that included measurements of pairs of relatives, e.g., siblings, parent and offspring, cousins, uncle and nephew, etc. Each of the pair had been measured once a year from one to 18 years of age. Correlations were calculated based on age-matched measurements (e.g., father's height at age 15 and his son's height at age 15). Multivariate analysis of the correlations found that degree of relatedness explained most of the variation in stature. That is, first degree relatives had higher correlations than second or third degree relatives. However, the effect of the common environment between first degree relatives, usually living in the same household, could not be separated from their genetic similarity. For example, between the ages of one and 15 years, correlations between siblings were always higher than those between parents and offspring. Theoretically, siblings, and parents and their offspring, share about 50 per cent more of their genes than the amount shared at random between any two unrelated members of a breeding population. Thus siblings and parents and their offspring should have approximately equal correlations in stature. However, the siblings lived together in the same households, which may have resulted in a 'commonality of environment' effect, increasing the value of sibling correlations. A similar pattern of familial correlations of growth was found by Susanne (1975), Russell (1976), and Mueller (1977), who also interpreted it as being due to the effect of a more similar environment for growth shared by siblings than by parent–offspring pairs.

Other studies show how familial correlations, although theoretically a measure of genetic similarity, are equally a measure of the environment. For instance, the power of the environment to influence the value of sibling correlations in size was demonstrated by Mueller & Pollitt (1982;

1983). They used data gathered by Dr Bacon Chow (Chow, 1974) from a study of the effects of nutritional supplementation of pregnant women on the subsequent growth of their offspring (Dr Chow died before analysis of the data could be completed). The study included measures of the prenatal and postnatal growth of siblings, who were living in a rural Taiwan village characterized by high rates of chronic undernutrition. Each woman in the study contributed two infant participants. During pregnancy with the first child the mother was untreated, while during pregnancy with the second child she was given either a high calorie supplement or a placebo. No supplement was given to the children directly, so any nutritional intervention relating to the growth of the children was mediated by the mother prenatally or during lactation. There were 108 pairs of siblings whose mothers received the high calorie supplement and 105 pairs of siblings whose mothers received the placebo. Correlations at birth in weight, length, head circumference, subscapular skinfold, and the index of weight/length^3 between siblings in the placebo group were significant and all were near the value of 0.50. Siblings in the supplemented group had birth size correlations that were 'unusually low and often insignificant' (Mueller & Pollitt, 1983, p. 11). The low correlations were due, presumably, to nutritional supplementation of the mother, which produced more favorable prenatal growth in the sibling exposed to the supplement. The differences in sibling correlations between the two groups virtually disappeared by age 2.5 years. Apparently, the maternal mediated effect of the high calorie supplement was limited to prenatal life and infancy. After weaning, the generally adverse nutritional environment of the village was a stronger influence on the growth of all the children.

Parent–child correlations in stature were studied by Martorell *et al.* (1977) for a sample of malnourished rural Guatemalan families. Correlations between mid-parent height (the average of the height of both parents) and child stature (or length) were obtained for children aged six months, one year, and then yearly up to age seven. The authors hypothesized that in this chronically malnourished sample, the stature of both the parents and the children would have been stunted. Furthermore, different degrees of malnutrition would have been experienced by different individuals. As a result, it was expected that the correlations between mid-parent stature and child stature for this sample would be lower than the values predicted from a genetic model, and lower than values from better nourished populations in developing countries. It was found that correlations for the Guatemalan sample did not differ from samples from the United States or northern Europe. It was possible to propose that variability in stature in the Guatemalan sample was as much a product of genetic influences as it was in the developed nations.

However, the authors rejected this proposal, and showed instead that socioeconomic and nutritional status were correlated across generations. That is, parents who had relatively better living conditions (e.g., housing, nutrition, etc.) when they were children were more likely to provide a better environment for their own children. Thus, environmental and genetic factors contributed to the parent–child correlation in stature, and it was not possible to quantify the unique contributions of either.

Familial correlation and heritability estimates for stature in a West African population were calculated by Roberts *et al.* (1978). The sample studied included the people of two villages in the Gambia, where traditional subsistence agriculture and rural lifestyles were practiced. The authors found that correlations for stature between husbands and wives were low and not statistically significant, indicating that there was no assortative mating for height. Correlations between parents and offspring, and between full siblings, for this sample were lower than those found for European or North American samples of middle to upper socioeconomic status. Moreover, the correlations between full siblings were lower than correlations between parents and children. The heritability for stature was estimated to be about 0.56 (1.00 being a perfect heritability and 0.00 indicating no heritability). To help put these African results in perspective, Byard *et al.* (1983) found, for a United States population, that sibling correlations were higher, generally, than parent–offspring correlations and that the heritability of stature was about 0.68. Roberts *et al.* suggested that their findings reflect a relatively larger environmental influence on stature than in the American or European studies. High rates of infant mortality (up to 50 per cent of newborns died by age five years), malaria, droughts and food shortages, and other 'rigours of the traditional way of life in West Africa' (p.23) all influenced the growth of the villagers. The authors emphasized that differences between generations, and between older and younger siblings, in the intensity of these environmental stresses would tend to lower familial correlations and heritability estimates.

Correlations in growth between adopted children and their adopted families

Garn *et al.* (1976; 1979) analyzed correlations in growth between adopted pairs of siblings, adopted and natural siblings, and parents and their adopted children. These contrasts provide another perspective of the relative importance of genes versus a common environment on growth. Correlations in fatness, measured as skinfolds, between biologically related parent-child pairs were higher ($r = 0.20$) than those for parents and their biologically unrelated adopted children ($r = 0.10$).

Between the ages of five and 18 years, biological siblings had higher correlations in fatness ($r = 0.27$) than unrelated adopted siblings ($r = 0.19$). In all cases, the correlation coefficients are significantly greater than zero, that is, no correlation.

The differences in magnitude between correlations based on biological and on social kinship indicate that genes play a role in the determination of fatness. However, the adopted sibling correlations and parent–adopted child correlations also were significant, even though a purely genetic model of growth determination predicts a zero correlation. The effect of living in a common family environment, called the cohabitational effect by Garn *et al.* (1976), may have been responsible for the higher than expected correlations in fatness between non-biologically related kin. Cohabitation also has been shown to result in significant correlations in energy intake, energy expenditure, serum vitamin levels, blood lipids, and stature in people with no biological relationship (Garn *et al.*, 1979). These correlations indicate, once again, the difficulty of identifying unique genetic effects on growth.

The effects of chromosomal abnormalities on growth

Although it may not be possible to separate the hereditary and environmental contributions to the phenotype, it is suspected that there are specific genes, or groups of genes, that can determine amounts of growth and rates of development. Studies of people with unusual karyotypes (the number and type of chromosomes inherited by an individual) provide evidence for such genes. Normal human karyotypes are 46,XY for males and 46,XX for women, 46 being the total number of chromosomes and X or Y being the types of sex chromosomes. Tanner *et al.* (1959) examined people with sex chromosome anomalies, including individuals with 47,XXY (Klinefelter's syndrome) and 45,X (Turner's syndrome) karyotypes. People with the 47,XXY condition are phenotypically males, and taller on average than normal 46,XY males, and people with the 45,X condition are phenotypically female, and much shorter, on average, than normal 46,XX females. Tanner *et al.* found that the body proportions (e.g., the ratio of leg length to stature) and rate of skeletal development of 47,XXY boys were like that of normal 46,XY boys. They also found that the rate of skeletal development of 45,X girls, up to puberty, was like that of normal 46,XX girls. Consequently, the authors concluded that genetic factors on the Y chromosome produce the male pattern of growth in body proportions and skeletal development.

Garn & Rohmann (1962) used a longitudinal sample of hand–wrist radiographs and dental radiographs to study ossification rate (number of bony centers present), ossification timing (age at the appearance of a center), and tooth calcification in siblings. The sample numbered 318 pairs of brother–brother, sister–sister, and brother–sister. Garn & Rohmann hypothesized that rates of skeletal and tooth development are genetically controlled and that some of these genes are linked to the sex chromosomes. They also proposed that pairs of sisters, who share the same paternal X chromosome, should have greater concordance in rates of development than pairs of brothers, who have only a 50 per cent chance of sharing the same maternal X chromosome, or brother–sister pairs, who share no paternal sex chromosomes. It was found that the correlation between pairs of sisters in skeletal and dental development (averaging about 0.52) was significantly greater than the correlation between pairs of brothers or brother–sister pairs (averaging about 0.35). Garn & Rohmann interpreted these correlations as evidence for X chromosome genetic control for rates of development.

The research strategy of searching for genetic determinants of growth by describing the size and development of individuals with sex chromosome anomalies has been used by many other investigators, as in a recent series of studies by Varrela and his colleagues. Varrela *et al.* (1984a) recorded 25 anthropometric measurements from 48 adult women with the 45,X karyotype (Turner's syndrome), 24 of their mothers and sisters and 95 control women to quantify the anthropometric differences associated with the chromosomal abnormality. The 45,X women were relatively smaller than their 46,XX mothers and sisters, and the control sample for all the length measurements (height, sitting height, leg length, etc.) and some breadth measurements (especially bitrochanteric), but relatively larger than their mothers, sisters, and the control sample in other measurements of breadth, circumference, and fatness. The authors believe that the cause of abnormal growth in the 45,X condition appears to be primarily genetic, though it is unclear if this is due to missing genes that contribute directly to growth in size, or to the impaired metabolic activity of 45,X cells. Varrela *et al.* cited other research showing that *in vitro* cultures of 45,X cells grow more slowly than cultures of normal 46,XX cells.

Varrela *et al.* (1984a) also found that the pattern of growth of 45,X girls may have been influenced by environmental factors. The subjects and controls of their sample were taller, and larger in several dimensions, compared with samples of 45,X women and normal women measured for earlier studies. The authors believe that this difference was due to improvements in living conditions since the time of the earlier studies, including higher socioeconomic status, better health care and nutritional status, and smaller family size, all of which are environmental changes that are known to be associated with increased growth. Another indication of the sensitivity of the growth of 45,X women to the environment is that Varrela *et al.*, and Brook *et al.* (1977) in another study, found that correlations in size between adult 45,X women and their mothers or fathers (about 0.49 and 0.56 in the studies cited) were significantly higher than correlations between 45,X women and their sisters (averaging 0.10 in the Varrela *et al.*, 1984a study). If the genetic cause of growth deficits in the 45,X condition is correct, the father to 45,X daughter correlations should be the lowest, since most affected daughters apparently lack the paternal, rather than the maternal, X chromosome in their karyotype. One estimate is that 80 per cent of the X chromosomes in 45,X females are of maternal origin (Sanger *et al.*, 1971). Given that the father's sex chromosome contribution is small, the significant correlation between father and 45,X daughter growth indicates an autosomal (non-sex chromosome) location of growth controlling genes or non-genetic influences on the size of the daughters. Furthermore, the correlations in growth between parents and their 45,X daughters were higher than those between 45,X girls and their sisters. This contrast may reflect differences in parental treatment of the normal daughter and the Turner's syndrome child. If so, the possibility exists that there are major environmental determinants of the growth of 45,X children.

Varrela (1984) measured the size of 29 adult men with Klinefelter's syndrome (47,XXY), finding that they were larger than normal controls in stature, arm length, leg length, and triceps and subscapular skinfold, but smaller than the normal controls in biacromial diameter, bideltoid breadth, wrist breadth, and most head dimensions. One interpretation of these findings is that there are genes on the X chromosome that control linear growth, though the direction of the growth control is not the same for all body parts. However, Varrela cites evidence that 47,XXY males are deficient in testosterone, but have normal plasma levels of growth hormone. It is known that during puberty, growth of the vertebrae and of the shoulders is dependent on androgenic hormones, such as testosterone, while leg growth is controlled more by growth hormone (Tanner *et al.* 1976a; Prader, 1984). Thus, the effect of the hormonal milieu associated with the extra X chromosome in Klinefelter's syndrome results in a relative increase in leg length and a relative decrease in biacromial (shoulder) breadth, which results in a more 'feminine' appearance compared with chromosomally normal males. The growth pattern of normal and affected males is not traceable to specific alleles for growth in size *per se*, rather it appears to be a hormone mediated effect.

In another study, Varrela *et al.* (1984b) measured body size of eight adult 46,XY women with complete testicular feminization. Although the 46,XY karyotype is usually associated with the male phenotype, some

46,XY embryos and fetuses lack a sensitivity of their tissues to androgenic hormones (Kelch *et al.*, 1972; Prader, 1984) and they develop into women, with normal female appearance. Varrela *et al.* found that in height, leg and arm length, and most other bony dimensions, the 46,XY women were larger than their mothers and sisters and a sample of normal controls. In body proportions the 46,XY females were not significantly different from normal females. The authors concluded that there are genes on the Y chromosome with a general size increasing effect, and this is one reason that normal men and 46,XY women are larger, on average, than normal women. They also suggested that the development of male body proportions requires a tissue sensitivity to androgenic hormones and not just the presence of Y-linked genes. For, in the case of testicular feminization, the body proportions of the 46,XY women are under the control of estrogens (or 'female' hormones), and these women achieve normal female body proportions.

To test the growth promoting effect of the Y chromosome further, Varrela & Alvesalo (1985) recorded 25 anthropometric measurements from seven men with the 47,XYY karyotype, and compared these with measurements from four normal male relatives and 42 normal control males. The 47,XYY males were significantly larger than the normal men in stature, sitting height, leg length, and bistyloid wrist breadth. The authors noted that the serum levels of growth promoting hormones in the 47,XYY males were about equal to the levels in normal 46,XY males, except for a 'slight' elevation in testosterone. Varrela & Alvesalo point out that no amount of testosterone alone could produce all of the growth differences between 47,XYY and normal males. As mentioned above, testosterone acts mainly to increase sitting height and shoulder width and does not strongly influence leg length. Moreover, the growth records of the 47,XYY males showed that they were all taller than average during childhood, before the pubertal increase in testosterone secretion occurred (Ratcliffe, 1976). This fact led the authors to conclude that the Y chromosome contains alleles that directly increase growth in linear dimensions, perhaps by a mechanism that increases cell proliferation.

These studies of sex chromosome abnormalities suggest an association between specific alleles located on the X and Y chromosomes and quantitative growth outcomes. However, the correlational studies of the growth of parents, children, and siblings discussed earlier suggest that the genetic control of growth is polygenic and autosomal. Taken together, these studies indicate that differences in growth that are attributable to genetic factors exist between individuals, and by extension, may exist between biological populations. Unfortunately, specific genes for growth have not yet been identified, though further advancement in chromosomal and molecular genetics could result in their discovery in the near future.

Endocrinology of growth

Hormones are organic substances synthesized in specific body tissues, often called endocrine glands. The glands secrete their hormones into the bloodstream, where they circulate to specific, and distant, sites of action. An example, discussed in the previous chapter, is the hormone cholecalciferol, or vitamin D_3, which is synthesized in the deep layers of the skin. It travels through the bloodstream to the liver and kidneys, where its biological activity is enhanced, and then to its sites of action, the intestine, where it promotes calcium absorption, and bone, where it regulates skeletal metabolism and bone growth. There are several major hormones that have an effect on growth and development, and these are discussed below. In addition, there are a group of substances known as growth factors that have effects on growth, both independently and interactively with each other and with hormones. Growth factors are synthesized by specific cells within a wide variety of body tissues. For instance, human liver and fibroblast cells produce substances known as the insulin-like growth factors (IGFs), which promote cell division in bone, muscle, and other tissue. IGF synthesis can be stimulated by growth hormone, from the pituitary gland, and both growth hormone and IGFs may need to be present to have an optimal influence on growth (Preece & Holder, 1982; Prader, 1984). D'Ercole & Underwood (1986) described how growth factors, like hormones, may be carried by the circulation to their sites of action (which they call the endocrine action), may act directly on the cells that synthesize them (an autocrine action), or may affect nearby cells in the same tissue (a paracrine action).

It is important to understand that the production and secretion of a hormone are only part of the process by which endocrine substances have an effect on the body. The target tissues for a hormone must be sensitive to its presence. Tissue sensitivity may be influenced by several factors, including the sex and age of an individual, the presence of biochemical receptors at the tissue level that bind with a hormone, and the production of 'secondary messengers,' so called because some hormones do not cross cell membranes and require intermediary substances to carry their 'message' into the target cells. The following account is limited to the more elementary aspects of the endocrinology of human growth. However, the reader should be aware that there are additional complexities of the endocrine system. Moreover, endocrinology is a very active area of research, and the current state of knowledge may be subject to substantial revision.

The major hormones of human growth and maturation

The actions and interactions of hormones and growth factors provide a system of fine control for the regulation of growth and development. A central feature of this system is the hypothalamic regulation of the pituitary gland. Figure 6.3 illustrates the location of the hypothalamus and pituitary at the base of the brain. Blood vessels directly connect the hypothalamus to the anterior pituitary gland, and allow for neurochemical communication from the hypothalamus to the pituitary. The hormones of the hypothalamus stimulate or inhibit the release of the

Figure 6.3. Diagram of the location of the hypothalamus and pituitary within the brain, and a schematic illustration of the target organs and tissues of the pituitary hormones. Abbreviations are: GnRH, gonadotropin releasing hormone; TRH, thyrotropin releasing hormone; PIF, prolactin release-inhibiting factor; CRF, adrenocorticotropin releasing factor; GHRH, growth hormone releasing hormone; GHRIH, growth hormone release-inhibiting hormone; GH, growth hormone; TSH, thyroid stimulating hormone; ACTH, adrenocorticotropic hormone; FSH, follicle stimulating hormone; LH, luteinizing hormone; MSH, melanocyte stimulating hormone (after Schally *et al.*, 1977).

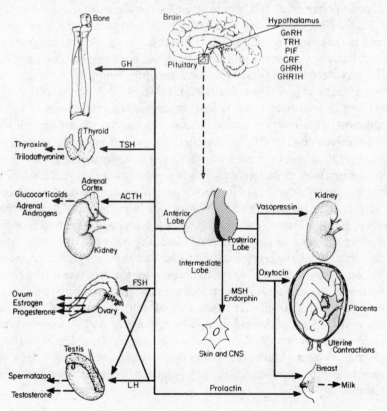

pituitary hormones, and these are released into the general circulation, where they act on specific target tissues throughout the body. Figure 6.3 also outlines the major hormones produced and secreted by each endocrine organ and indicates the target tissues of the pituitary hormones. The following discussion is confined to those hormones involved in growth and maturation.

Thyroid hormones

Thyrotropin releasing hormone (TRH), secreted by the hypothalamus, stimulates the release from the pituitary of thyroid stimulating hormone (TSH). The pituitary TSH acts on the thyroid gland to promote the release of two metabolically active hormones, thyroxine and triiodothyronine. A negative feedback relationship controls the release of TSH and the two thyroid hormones into the bloodstream. In Figure 6.4 three negative feedback loops are illustrated showing how thyroid activity is controlled. This model may be applied, generally, to the other hypothalamic–pituitary–target tissue hormones. The ultrashort feedback loop involves auto-feedback of TRH within the hypothalamus.

Figure 6.4. Feedback circuits for the control of the hypothalamic–pituitary–thyroid system.

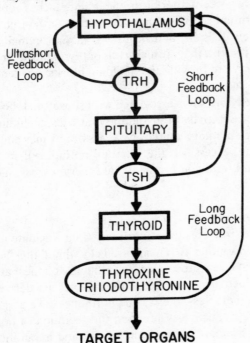

In this case, rising levels of TRH within the hypothalamus may suppress the activity of TRH secretory cells directly. A second form of feedback control involves rising systemic levels of pituitary TSH, which may decrease the release of TRH through a short feedback loop. Finally, an increase in the blood levels of thyroxine and triiodothyronine may suppress TRH secretion through a long feedback loop. All three avenues of feedback control may work simultaneously to 'fine-tune' the level of thyroid hormones in the bloodstream.

A fine level of control is needed, since thyroxine and triiodothyronine have powerful metabolic actions. Thyroid hormones are needed for normal growth in stature, the development of normal body proportions, formation of bone from cartilage, and formation of the teeth. A deficiency of these hormones (hypothyroidism) during infancy and childhood results in growth retardation and mental impairment, and in the extreme case the child suffers from a form of dwarfism called cretinism. Thyroid hormones seem to have an important role in the maturation of brain enzyme systems and myelination, the covering of nerve fibers with a fatty insulation which speeds up the transmission of nerve impulses. This explains, in part, why hypothyrodisim in infants results in mental impairment. Sizonenko & Aubert (1986) reported that thyroxine has been detected in the human fetus at 78 days of age, and serum levels rise, generally, throughout the prenatal period. By two weeks after birth, the infant reaches adult levels of serum thyroid hormone activity. The exact nature of the role thyroid hormones play in cell division, and in growth in size, remains to be determined, though it seems that normal levels of thyroxine favor protein synthesis and thus serve to provide the 'building blocks' needed for the growth of all body tissues.

Adult-onset hypothyroidism is associated with slower metabolic rate, resulting in a decreased activity of involuntary muscles, including heart and gut muscles, in voluntary muscle weakness, and in weight gain. Hyperthyroidism (an excess of the thyroid hormones) speeds up metabolic activity and results in fast heart rate, nervousness, increased appetite, and weight loss.

Gonadal hormones

The pituitary secretes two hormones that regulate gonadal activity: luteinizing hormone (LH), and follicle stimulating hormone (FSH). They were first discovered to have an influence on the maturation of the female reproductive system, hence their names are derived from ovarian functions. Nonetheless, these same hormones have a significant effect on the male reproductive system and the secretion of hormones from the testes that have an influence on growth and maturation. The pituitary gonadotropins have but a single hypothalamic releasing hormone (Schally *et al.*, 1977), called gonadotropin releasing hormone (GnRH).

In women, FSH and LH stimulate the growth of the ovaries and the release of ovarian hormones. The ovarian hormones are called, collectively, the estrogens. In men, FSH promotes the development of the seminiferous tubules and initiates the production of spermatozoa. LH stimulates secretion of hormones from the testes, collectively called the androgens. Androgens are required to complete the formation of mature spermatozoa and these hormones also have an influence on the growth of bone, muscle, fat, and other tissues. In both men and women, the serum levels of androgens or estrogens regulate the secretion of GnRH, LH, and FSH through a, generally, negative feedback relationship (similar to that for the thyroid hormones). In women, however, serum estrogen levels exert a positive feedback influence on the hypothalamus and pituitary during the pre-ovulatory phase of the menstrual cycle, leading to an LH surge at mid-cycle that results in ovulation.

Gonadotropins, estrogens, and androgens have major influences on human development. LH and FSH may be detected in fetal pituitary tissue as early as ten weeks of age (Sizonenko & Aubert, 1986). In the male fetus, LH and FSH probably stimulate the undifferentiated gonad to develop into the testis and produce and secrete androgens, especially testosterone. The source of the prenatal testosterone may not be from fetal tissues entirely. It is known that a chromosomal male embryo lacking the hypothalamus or pituitary develops into a phenotypically male fetus, though with somewhat underdeveloped penis and testes. A placental hormone, human chorionic gonadotropin, is likely to be capable of stimulating differentiation of the testis and production of testosterone. This placental hormone may play a role in the development of the normal fetus as well. The role of LH and FSH in the development of the female embryo is not well understood.

At birth, and up to about two years of age, serum levels of LH, FSH, and gonadal hormones are higher than at any time prior to the onset of puberty. Noting this fact, Grumbach *et al.* (1974) suggest that the sensitivity of negative feedback in the system of hypothalamic–pituitary–gonadal regulation is not well developed in the infant. The delay in maturation of the feedback system may be related to the growth and development of the child during infancy and later in life. The relatively high levels of estrogen and androgens in the bloodstream of the infant are correlated with the rapid velocity of physical growth, neurological development, motor control, and cognitive advancement that occurs during infancy. The rate of production of gonadal hormones, and growth velocity, fall to relatively low childhood levels by about two years of age, when the negative feedback control becomes highly sensitive. Grumbach *et al.* also note that at adolescence, a new positive feedback control develops, in which rising levels of gonadal hormone secretion result in an increased secretion of GnRH, LH, and FSH. During adolescence the rate of growth increases rapidly (i.e., the growth spurt occurs). Prader (1984) and Preece (1986) find that the rising levels of testosterone and estradiol are significantly correlated with growth rate during puberty, and they suggest that increased secretion of these hormones is likely to be one cause of the adolescent growth spurt.

Adrenal hormones

As was shown in Chapter 2 of this book, most mammals progress from infancy to adulthood without any intervening period of childhood development. Human development interposes the childhood growth period between infancy and sexual maturation. The relative inactivity of the hypothalamic–pituitary–gonadal system during childhood is an evolutionary novelty, delaying sexual and mental maturation, and, possibly, helping to maintain the slow rate of physical growth, and the high rate of learning, that characterize human childhood. The only remarkable endocrine event of this period is adrenarche, which was discussed in Chapter 1, and shown to be related, perhaps, to the mid-growth spurt in height velocity experienced by some children. One study indicates that, like childhood in humans, adrenarche is an evolutionarily novel event, found only in apes and people. Cutler *et al.* (1978) measured the plasma concentration of three adrenal androgens, dehydroepiandosterone (DHA), dehydroepiandosterone sulfate (DHAS), and delta^4-androstenedione (D^4), before and after sexual maturation in 13 mammalian species including rodents (rat, guinea pig, and hamster), domestic animals (rabbit, dog, sheep, pig, goat, horse, and cow), primates (rhesus monkey and chimpanzee), and the chicken. The plasma concentrations of DHA, DHAS, and D^4 were significantly higher in sexually mature primate species than in any of the other animals. Rhesus monkeys aged one to three years old, and not sexually mature, had the same high concentrations of all three adrenal androgens as older, sexually mature, monkeys. In contrast, chimpanzees seven years old or older had adrenal androgen concentrations that were on average 4.7 times greater than those for chimpanzees less than four years old. Thus among the animals examined so far, the chimpanzee and the human being are the only species that show adrenarche.

The findings of Cutler *et al.* may be used to develop some speculative hypotheses about the evolution of the human pattern of growth. The data show that relatively high levels of adrenal androgen production and secretion appear to be a primate characteristic. However, there is variability between primate species in the relation of age and adrenal activity. Rhesus monkeys have relatively high plasma concentrations of DHA, DHAS, and D^4 at all ages after birth and, compared with chimpanzees and people, rhesus monkeys sexually mature at a relatively early age. Chimpanzees and humans have relatively low serum levels of adrenal androgens prior to adrenarche. Moreover, chimpanzees and humans have a relatively long delay in the onset of sexual maturation. In the human being, the delay is so protracted that the childhood period of growth becomes a major feature of the total pattern of development. Perhaps the evolution of human growth may be explained, at least in part, by the mechanisms that control the activity of the adrenal gland and the production of its androgenic hormones.

At present, the mechanism controlling the production of adrenal androgens in human beings is not well understood. The cortex of the adrenal produces two classes of hormones; glucocorticoids, which are involved in the body's ability to maintain homeostasis when faced with physical or emotional stress, and androgens. Adrenal androgens are produced by the *zona reticularis* of the adrenal cortex, which is relatively large and active during fetal life, but undergoes involution after birth. Adrenal androgen levels are low throughout childhood until adrenarche, about age six to nine years, when secretions begin to increase steadily until a plateau is reached late in the fourth decade of life (Weirman & Crowley, 1986).

The control of adrenal androgen production may involve the activity of glucocorticoids, and their pituitary stimulating hormone called adrenocorticotropic hormone (ACTH). Anderson (1980) proposed that rising levels of the glucocorticoids during the late fetal period act to suppress adrenal androgen production. Years later, stimulation by ACTH overcomes the inhibiting action of the glucocorticoids and produces adrenarche.

Compounding the problem of understanding the evolutionary importance of adrenarche is the fact that its physiological function is not known completely. Katz *et al.* (1985) found a positive correlation between levels of adrenal androgens, skeletal maturation, and fatness in adolescent boys that was independent of their serum levels of gonadal testosterone. In a related study, Zemel & Katz (1986) calculated the statistical contribution of testosterone and the major adrenal androgen (dehydroepiandosterone sulfate) to growth velocity in height for a sample of 181 male adolescents. From a regression analysis, it was found that both hormones had a significant effect on height velocity. However, the relative influence of

testosterone was about five times greater than that of the adrenal androgen. Hediger & Katz (1986) found that increased adrenal androgen levels
in skeletally mature young women were correlated positively with fatness
and associated with a concentration of subcutaneous fat on the trunk of
the body. Taken together, this series of studies by Katz and his colleagues
suggest that adrenarche may be one of the endocrine events that promotes
maturation and determines, in part, adult body proportions and body
composition.

Adrenarche and the onset of puberty

At one time, it was proposed that adrenarche might be a step in
the process that results in gonadarche, the maturation of the reproductive
system. Adrenal androgens were thought to 'prime' the gonads and
stimulate their maturation. However, Weirman & Crowley (1986)
reviewed several studies that demonstrate the physiological independence of the two events. That is, children with early or late adrenarche do
not have a correspondingly early or late gonadarche. In fact, some
children have no appreciable adrenarche, but experience a perfectly
normal sexual maturation, at the appropriate age, during adolescence.
Worthman (1986) analyzed endocrine changes during childhood and
adolescence for 54 girls and 48 boys of the Kikuyu tribe of Kenya, Africa.
The onset of puberty is relatively late in this population, girls achieving
menarche at a median age of 15.9 years. From gonadal hormone profiles,
girls enter puberty at a median age of 13.0 years and boys at a median age
of 13.5 years. These ages are about two years later than those for children
living in the United States. Despite the delay in the onset of puberty, the
Kikuyu children experience adrenarche at about the same age as children
from the United States, which does not support the notion that
adrenarche and gonadarche are causally linked.

Hypothalamic and pituitary regulation of puberty

The first endocrine signs of puberty in boys are increases in the
nighttime secretion of gonadotropin releasing hormone (GnRH) by the
hypothalamus, LH by the pituitary, and testosterone by the testes. Boyar
et al. (1974) and Judd *et al.* (1977) measured the serum levels of these
hormones in normal boys by drawing blood every 20 to 30 minutes over a
24-hour period. In boys with an average bone age of eight 'years' or older,
hormone levels began to show an increase during the night, while the boys
were asleep. Serum LH levels increased about four-fold and testosterone
levels increased about five-fold over prepubertal base-line values. During
the daytime, all endocrine levels remained at base-line values. These
nighttime episodes of GnRH, LH, and testosterone release preceded

development of primary or secondary sexual characteristics. Weitzman *et al.* (1975) found that by middle to late adolescence, when the secondary sexual characteristics were well on their way towards maturity, the secretion of LH occurred in 'pulses' during the day as well as during the night. Testosterone levels remained relatively high at all hours in these post-pubertal boys. This endocrine pattern is also found for adult men.

Boyar *et al.* (1976) did not find a nighttime increase in LH or FSH in adolescent girls, but did find a peak in serum levels at about 3.00 p.m. Jakacki *et al.* (1982) found that one endocrinologically normal 11.8-year old girl, of two girls studied, had significant nighttime LH secretory pulses. Whether secretion occurred by day or by night, both studies found that, as for boys, a pulsatile secretion of GnRH and LH is necessary to initiate and maintain normal gonadal secretions of estrogens (or androgens in boys). Experimental studies show that a constant infusion, or single daily doses, of LH administered to LH-deficient children, monkeys, or rats will not promote gonadal maturation, but doses administered in a pulsatile fashion will do so (Marshall & Kelch, 1986).

These findings suggested to some researchers (e.g. Grumbach *et al.*, 1974) that the onset of nighttime and/or pulsatile secretions of LH and gonadal hormones marked the change from the childhood pattern of hypothalamic–pituitary regulation of the gonads to the pubertal pattern of reproductive system control. That is, during childhood relatively low levels of circulating gonadal hormones acted to inhibit the secretion of releasing factors from the hypothalamus and gonadal stimulating hormones from the pituitary. At puberty one of two changes was believed to occur. Either the feedback system was reset, so that much higher levels of gonadal hormones were required to inhibit the hypothalamus, or gonadal hormones acted to stimulate the hypothalamus and pituitary to release even more gonadotropic hormones. More recently, Jakacki *et al.* (1982) used a sensitive assay method to measure the gonadotropin secretions of 15 prepubertal children with bone ages less than 10 'years.' Blood samples were taken every 20 minutes for periods of three to 11 hours. Nocturnal levels of LH and FSH were higher in all children and a relatively low pulsatile secretion of LH was detected in eight of the children, some with bone ages as young as five 'years.' The authors concluded that 'puberty is heralded by the amplification, rather than the initiation, of a circadian pattern of gonadotropin secretion' (p. 457). Thus, it appears that the hypothalamic–pituitary–gonadal system does not change its feedback relationships, rather it seems that low levels of GnRH, LH, FSH, and gonadal hormone secretion during childhood represent the inhibition of the reproductive endocrine system by some, as yet unknown, mechanism. At about eight to 10 years of age, the inhibiting mechanism relinquishes control and the hypothalamic–pituitary–gonadal system starts to approach mature levels of operation.

The pineal gland, melatonin, and puberty

Although the mechanism by which GnRH, LH, and gonadal hormone secretion is inhibited during childhood and stimulated with the onset of adolescence is not known, some studies have indirectly indicated that the pineal gland may be involved. The pineal gland is located roughly at the center of the brain, near the inferior dorsal margin of the third ventricle, and contains or produces several compounds. Melatonin is the only one of these compounds for which a biological function is known (Tamarkin *et al.*, 1985). Wurtman & Axelrod (1965) described how cells in the pineal gland transduce the photic information carried by the nervous system into the endocrine product melatonin, which is secreted into the general circulation. In seasonally breeding animals, changes in day length during the year correlate with levels of melatonin in circulation and the onset or suppression of gonadal maturation. In relation to human puberty, Wurtman & Axelrod reported that for many years there had been speculation that high levels of melatonin in children might serve to suppress reproductive maturation. For instance, pineal tumors have long been known to be associated with precocious puberty in boys. Such children have an unusually early puberty, usually occurring before the age of eight years, and also have an abnormal pattern of growth. They are relatively tall during childhood, but this is due to an early pubertal growth spurt. The early spurt means that growth in height ends at a young age, often before the teenage years begin, which means that as adults these individuals are relatively short.

Supporting the hypothesis that the pineal gland may help regulate puberty, Silman *et al.* (1979) found that in a sample of boys, daytime levels of circulating melatonin were relatively high in early childhood and dropped to relatively low levels in early to mid-adolescence. No change in daytime melatonin levels was found for girls or for nighttime levels for either sex. Lenko *et al.* (1981) found no age relationship to daytime melatonin secretion for boys or girls. Further confusion arose when Waldhauser *et al.* (1984) discovered a significant age related decrease in nighttime melatonin levels, but not in daytime levels, for a sample of hospitalized boys and girls. Tetsuo *et al.* (1982) reported no age change in total melatonin secretion, but did note that as expressed per kilogram of body weight, melatonin concentration decreased as children grew larger. This suggested that at some threshold level of body size, melatonin concentration would be so dilute as no longer to suppress reproductive maturation. Tamarkin *et al.* (1985) found no evidence of age-related changes in serum melatonin levels in normal children or those with precocious puberty. In contrast to the previous studies, which were based on data derived from single blood samples taken during the day or night, Tamarkin *et al.* collected blood at two to three hour intervals for 24 hours. This methodology may be the only one capable of revealing the daily profile of melatonin, or any hormone, secreted in an episodic pattern. Tamarkin *et al.* noted that, to date, only the amount of melatonin secretion has been measured and not temporal changes in the rhythm of melatonin secretion, which may serve as the biological regulator of sexual maturation.

Effects of gonadal hormones on growth and maturation at puberty

Although the causes of the onset of puberty are not known, the consequences of hypothalamic maturation for the gonads and for physical development are well documented. Preece (1986) pooled the results of several studies and found that, for girls, the amplification of pulsatile secretion of GnRH results in an increase of serum levels of LH and estradiol during puberty; for example, estradiol levels rise from about 50 picomoles/liter at age seven years to over 400 pmol/l by age 15 years. Preece also analyzed studies of children from whom both growth and hormone measurements were taken, to verify relationships between levels of gonadal hormones and physical maturation that had been long suspected but never proven. It was found that the increase in estradiol levels is highly correlated with secondary sexual development, for example, maturation of the breasts. In an earlier study, Preece *et al.* (1984) discovered a similar high correlation between testosterone levels and genitalia development in adolescent boys. Using Tanner's (1962) system of rating genital maturation stages, Preece *et al.* found that at genitalia stage 2, when the first signs of enlargement of the penis and testes are clinically detectable, the serum concentration of testosterone averaged about 7.0 nanomoles/liter. At genitalia stage 5, when the penis and scrotum have an adult size and appearance, serum testosterone concentration averaged about 19 nmol/l.

Prader (1984) and Preece *et al.* (1984) reviewed studies that show that a pubertal increase in testosterone secretion is needed for a normal growth spurt to take place in boys. Of course, normal levels of other growth promoting hormones, such as thyroxine and growth hormone, must be present for a normal growth spurt to occur. For about two years before the peak height velocity (PHV) of the adolescent growth spurt, rising testosterone levels are positively and significantly correlated with increasing rates of growth in height. For at least three years after PHV, rising testosterone levels are negatively and significantly correlated with rate of height growth. Preece (1986) states that these findings suggest that the principal effect of testosterone in early adolescence is to stimulate bone growth, but in later adolescence its main action is to promote epiphyseal fusion, which slows growth. It is also likely that in the later stages of puberty, testosterone exerts an inhibiting effect directly on the growth of cartilage and bone cells, since boys with higher levels of testosterone terminate their growth spurt more quickly than boys with lower levels of testosterone (Preece *et al.* 1984). Prader (1984) reviewed endocrine studies of the control of the adolescent height growth spurt in girls. Girls produce androgens, in the adrenals and by conversion of ovarian estrogens to androgens, which have been shown to stimulate bone growth. However, girls lacking these androgens or lacking tissue sensitivity to androgens, such as girls with testicular feminization (see discussion of such cases in the genetics section of this chapter), experience a normal growth spurt. From this, Prader concluded that, in otherwise normal girls, the high levels of estradiol secreted during female puberty are directly responsible for the growth spurt. Presumably, estradiol acts like testosterone to stimulate bone growth in early puberty and to suppress cell division and facilitate epiphyseal closure in later puberty.

Growth hormone and growth factors
Throughout prenatal and postnatal life the availability of growth hormone and other growth factors are necessary to maintain and promote physical growth and development. Growth hormone (GH) is synthesized and secreted by the anterior pituitary gland. A hypothalamic hormone, growth hormone releasing factor (GHRF), stimulates the synthesis of GH and, along with other agents, causes the release of GH into the bloodstream (Barinaga *et al.*, 1985). Another hypothalamic hormone, growth hormone release-inhibiting hormone, also called somatostatin, has an antisecretory effect on GH. At specific locations in the body, somatostatin also inhibits the secretion of TSH, glugagon, insulin, and several digestive acids and enzymes (Schally *et al.*, 1977). Experimental research with rats indicates that secretion of somatostatin and GHRF are inversely correlated, suggesting that the feedback control of GH secretion is maintained within the hypothalamus (Plotsky & Vale, 1985).

Unlike the other pituitary hormones so far discussed, GH does not seem to affect a single target tissue, but appears to have a general growth promoting effect throughout the body. Early research by Cheek (1968) found that GH is needed for the body to retain nitrogen, sodium, chlorine, potassium, phosphorous, calcium, and other elements that make up body tissue. GH is also needed for muscle cell division. Isaksson *et al.* (1982) demonstrated experimentally that GH stimulates long bone growth directly. The authors injected GH into the cartilage growth plate of the proximal tibia of rats, in which the pituitary gland had been surgically removed. Rat tibia receiving the GH injection grew significantly more than rat tibia receiving a saline injection. Green and his colleagues (Morikawa *et al.*1982; Nixon & Green,1984; Green *et al*. 1985) found that GH promotes the differentiation of pre-adipose germ cells into adipose cells. Given the widespread action of GH in the body, the name 'growth hormone' is an accurate description of the function of this endocrine product.

Another class of growth promoting substances are called the somatomedins or insulin-like growth factors (IGFs). There is currently some debate about the relative roles that GH and the IGFs play in cell division and growth. The IGFs are similar to insulin in molecular structure and in biological action (Preece, 1986). Insulin, a hormone produced and secreted by cells of the pancreas, stimulates protein synthesis and the growth of cartilage cells. It is well known that GH acts on the pancreas to increase the synthesis of insulin, and also increases the serum levels of the IGFs. Although it was once thought that the origin of IGFs was from cells located in the liver, it is now known that a wide variety of cells produce IGFs (Clemmons & Van Wyk, 1984). In humans there are two major types of IGF: IGF–1, which may be the type that, in concert with GH, regulates postnatal growth, and IGF–2, which appears to be the type that controls some aspects of prenatal growth and is stimulated by placental lactogen, a hormone similar to GH and prolactin (Sizonenko & Aubert, 1986).

Preece & Holder (1982) reviewed experimental and clinical studies of the role of GH and the IGFs on growth. They argued that the evidence suggested a cascade effect: hypothalamic GHRF stimulating the release of pituitary GH, which circulates in the bloodstream stimulating the production of the IGFs at the tissue level, where they have autocrine and paracrine actions to promote cell division. Preece & Holder cite several endocrine pathologies as evidence for the role of IGFs as the direct growth promoting substance, in particular the conditions Laron dwarfism and acromegaly. Dwarfism may be due to low levels of, or an absence of, GH, but Laron dwarfism results from low levels of IGFs despite normal or high levels of GH. Acromegaly is a condition of abnormal growth characterized by a slow, but continuous increase in size of the bones of the face, hands, and feet throughout life, that often leads to gross disfigurement in adulthood. Levels of GH and IGFs are both above normal in acromegalic patients; however, there is a higher correlation between IGFs and the clinical progress of the disease.

Clearly, the IGFs have an important role in normal and abnormal growth. However, recent studies confirm that GH alone can stimulate cellular growth. As mentioned above, studies of living animals and cultured cells find that GH administration appears to promote the differentiation of cartilage cells into bone cells, and pre-adipose cells into adipose cells (Isaksson *et al.*, 1982; Morikawa *et al.*, 1982). A 'dual effector' system was proposed by Green *et al.* (1985) to model the coordinated actions of GH and the IGFs. In this model the growth of tissues occurs in two stages, differentiation of cells from their germ layer precursors and, then, cell growth by hyperplasia and hypertrophy. The cell growth must be selective for only the newly differentiated cells, since a general response of growth by all cells in a tissue would result in uncontrolled hyperplasia, leading, perhaps, to cancerous growth. In the 'dual effector' system, the differentiation of cells is controlled by GH and the selective multiplication of young differentiated cells controlled by specific IGFs.

Zezulak & Green (1986) found that this model accounts for the growth of adipose cells in culture. GH stimulated the differentiation of pre-adipocyte germ cells into young adipocytes, and IGF–1 promoted the multiplication of these young cells by mitotic division. Nilsson *et al.* (1986) demonstrated that the 'dual effector' model also applies to bone growth. GH was found to stimulate the differentiation of pre-chondrocyte stem cells, in the growth plates of the rat tibia, into chondrocytes (cells that will form cartilage and, eventually, bone). The presence of IGF–1 was concentrated in the young chondrocytes, where it presumably stimulated the multiplication of those cells by mitosis. It appears that IGF–1 is produced locally, within the growth plate, and that it exerts its function, 'through paracrine or autocrine mechanisms' (p. 233).

Human studies must be conducted before these results in support of the 'dual effector' model can be generalized to our species. Furthermore, since several elements of the 'dual effector' model overlap with the endocrine cascade model of Preece & Holder, there is a need for further research to clarify the independent and interactive roles that GH and IGFs play in human growth.

GH, IGFs, and rate of growth in normal children

Sizonenko & Aubert (1986) reviewed research showing that GH levels rise continuously throughout fetal life, peaking at the maximum lifetime level at about 35 to 40 weeks of gestation. The high value of GH in serum at this time may have two causes. First, the fetus uses placental lactogen as the 'growth hormone,' rather than GH. Second, the negative feedback system for the regulation of GH secretion is not fully functional during fetal life. After birth, GH levels in serum decrease, so that at about one month of age they reach adult values, and at about three months of age the GH regulatory system seems to mature. Throughout the rest of life, GH is secreted in a pulsatile fashion, primarily at night. However, physical and emotional stresses are capable of producing episodes of GH secretion at any time of day. Preece *et al.* (1984) found for boys that levels of IGFs, and possibly GH, were higher during adolescence than during childhood or adulthood. In fact, IGF levels were highest at the time of peak growth velocity in height; however, the relationship of IGFs to growth rate was not statistically significant. Rosenfeld *et* al. (1983) measured IGF levels during puberty in boys and girls, finding that peak IGF values usually occurred about a year after a child achieved peak height velocity. From these studies, it seems that GH and IGFs play, at best, a limited role in the adolescent growth spurt. Prader (1984) argued that normal levels of GH and IGFs are needed to maintain normal adolescent growth, but that the pubertal growth spurt is more directly related to the marked increase in the secretion of testosterone or estradiol, and not GH.

Other growth factors

In addition to the IGFs, which appear to be regulated, in part, by GH release, there is ever increasing recognition of other growth factors which are specific for different tissues. Studies of the action of these growth factors was reviewed by D'Ercole & Underwood (1986). Epidermal growth factor and fibroblast growth factor appear to act like IGFs, in that they stimulate cell division of newly differentiated epidermal or fibroblast cells. Platelet-derived growth factor is released from platelets during blood coagulation, and may have an important function in response to injury and during cell differentiation. Bone-derived growth factor appears to be capable of stimulating cell division in cartilage and bone cells. Nerve growth factor may have little or no mitogenic (cell division) influence, but seems to be needed to differentiate nerve cells. There are other known or postulated growth factors. How they all act and interact during the growth of the individual person, or any experimental animal, is not well understood, nor is it known if they are unique substances or members of a family of closely related autocrine and paracrine factors. Other pituitary hormones, such as prolactin, melanocyte stimulating hormone, vasopressin, oxytocin, and their hypothalamic releasing or inhibiting factors are necessary for normal metabolic activity, the maintenance of the placenta and fetus, birth, and other life sustaining processes. Non-pituitary hormones play a similarly vital role. In a broad sense, therefore, these hormones are essential for growth, but a detailed discussion of them is outside the scope of this book.

Endocrine mediation of genes and the environment

The genetic–environmental interactions that influence the course of growth and development of the individual may require the endocrine system to produce their effects. For instance, it is often stated that the average short stature of the Mbuti and other pygmy peoples of Central Africa is of genetic origin. An argument by Hiernaux (1974) to support this notion, based on natural selection for optimal heat loss in the forested environment of the Mbuti, was presented in the previous chapter. However, to state simply that the size of the pygmies, or any other population, is due to genes is unsatisfactory, for such a statement does not involve any mechanism for growth control.

In the case of the pygmies several endocrine mechanisms have been suggested. Rimoin *et al.* (1968) and Merimee *et al.* (1968) found that after administration of a provocative stimulus, serum levels of growth hormone and IGF–2 were normal in the sample of Central African pygmies they studied. However, Merimee *et al.* (1981) found that IGF–1 levels were significantly lower in serum samples drawn from pygmies than from taller peoples living near the pygmies. A nutritional cause for the low IGF–1 levels (see below) and for short stature could not be demonstrated. Rather, it seems that a genetic defect in the cellular mechanisms for the production or release of IGF–1 is the cause of the short stature of the pygmies. Eigenmann *et al.* (1984) found that the small size of toy and miniature poodles is due to a defect of IGF–1 production. Both types of poodles have normal serum concentrations of GH and IGF–2, but significantly lower concentrations of IGF–1 than found in standard poodles. The smaller toy poodles have levels of IGF–1 lower than the miniature poodles. The endocrine similarities between the poodle models and the human pygmy situation with respect to IGF–1, add support to a genetic hypothesis for the determination of pygmy stature.

Schwartz *et al.* (under review) studied the endocrine status of another short stature population, the Mountain Ok of Papua New Guinea. The serum levels of GH, IGF–1, and IGF–2 all were found to be in the normal range. The mean height of men (152.7 cm) and women (146.7 cm) in the Mountain Ok sample ($n = 150$ subjects) could not be associated with low values of any of the growth promoting hormones. The authors point out that this underscores the uniqueness of the low IGF–1 levels associated with short stature of the African pygmies. It is more likely that the small size of the Mountain Ok has a nutritional cause. Lourie *et al.* (1986) surveyed 79 per cent of the Mountain Ok population for a variety of anthropometric indicators of health and nutritional status. The authors find much evidence for chronic calorie undernutrition, for instance, relatively small skinfolds and arm circumferences in people of all ages.

Another indication of a nutritional cause for short stature is that Mountain Ok living in villages within a day's walk of a mining town (gold and copper are mined in the region) are significantly taller, heavier, and fatter than Mountain Ok living in more remote villages. Opportunities for wage labor and the greater availability of food in mining towns are likely reasons for the differences in growth that were found.

Undernutrition during childhood results in slow growth, delayed maturation, and, if prolonged into adolescence, short stature in adulthood. Pimstone *et al.* (1968) found that GH levels were higher than normal in children suffering from kwashiorkor, a severe form of protein undernutrition, and from marasmus–kwashiorkor, a severe form of protein–energy malnutrition. Both groups of children were significantly shorter and lighter than expected for their ages. Upon recovery in the hospital, GH levels remained high when calories only were added to the diet, but fell towards normal values when protein and calories were added.

It may seem paradoxical that severely malnourished children should have high levels of GH but delayed growth. The elevation of GH may be due to impaired release of IGFs and the lack of negative feedback of rising levels of IGFs on further GH release. Grant *et al.* (1973) showed that children with marasmus–kwashiorkor have relatively high GH levels and relatively low IGF levels. Three boys and two girls, aged 11 to 32 months old, were studied following admission to a South African hospital. After recovery for nine days in hospital, IGF levels had risen, while GH levels had fallen in the patients. The authors speculated that starvation reduces IGF secretion and activity as a means of saving body stores of protein for the maintenance and repair of tissues, at the expense of cellular growth. These findings were repeated in a study by Hintz *et al.* (1978) of malnourished children in Thailand. Twenty-seven children, aged eight to 60 months old, were admitted to hospital, 24 with marasmus and three with kwashiorkor. Twenty-one healthy, well-nourished children in hospital for minor surgical procedures were studied as controls. The IGF levels of the malnourished children were significantly lower than those of the controls at days two and eight after admission to the hospital. By days 29 and 50, the IGF levels of the malnourished children were not significantly different from the controls. There was an inverse relationship between GH and IGF levels in the malnourished group, that is, GH values fell as IGF values increased. Hintz *et al.* suggested that some inhibiting factor must block the synthesis and release of IGF, and they believe they found evidence for such a factor in samples of the serum of 10 out of the 27 patients. To date, the source and biochemical nature of this IGF 'inhibitor' has not been described.

In a controlled experimental study, Clemmons *et al.* (1981) found that fasting for 48 hours significantly decreased IGF–1 levels. The subjects were seven adult men suffering from obesity, who had volunteered to fast for 10 days as part of a weight reduction experiment. In another study, Isley *et al.* (1983) placed five normal weight people (three women and two men) on a total fast, an energy deficient diet, and a protein deficient diet. The subjects were on each type of diet for five days, followed by five more days of recovery, and then a further five days on a diet until all subjects had experienced each type of diet. The authors found that IGF–1 levels in serum dropped on each type of diet, and fell steadily each day, so that they reached their lowest values after five days. IGF–1 levels rapidly returned to normal values with refeeding.

Each of these studies of endocrine-nutrition interactions shows that, in addition to lack of energy, protein, and other nutrients needed for increases in cell size and number, low IGF levels in the undernourished state probably contribute to slow growth. These studies also suggest that in humans, GH acting alone cannot stimulate growth; both GH and IGFs must be available. Prader *et al.* (1963) described several clinical cases of prepubertal children who experienced a dramatic increase in growth velocity during recovery from acute undernutrition or digestive tract diseases (e.g., coeliac disease) that block nutrient absorption. This phenomenon was confirmed in many subsequent studies and is known, generally, as 'catch-up' growth. The physiological mechanism controlling catch-up growth in these young children is not known. Part of the mechanism, however, may be the rapid rise in IGF secretion upon recovery, which could cause a sudden proliferation of bone, muscle, and adipose tissue throughout the body.

Psychological correlates of endocrinology

The quality of the emotional and psychological environment in which a child lives is known to influence his or her growth. Many researchers believe that the endocrine system mediates the relationship between psychological factors and physical growth. For instance, in a review of his clinical experience with psychosocial dwarfism, Rappaport (1984) stated that 'the most consistent biological finding was the decrease of circulating somatomedin [IGF] activity' (p. 44). In contrast, other clinicians believe that neglect and undernutrition are the direct causes of psychosocial dwarfism (Sills, 1978).

Bakwin (1942) and Spitz (1945) were among the first to investigate carefully the causes of poor growth experienced by many children confined to foundling homes and other institutions. Spitz compared the development of infants in a foundling home with infants raised in the nursery of a penal institution for delinquent girls. Inmates of the latter prison were the natural mothers of the infants. Both institutions provided an acceptable standard of housing, sanitation, medical care, and diet for the infants. The children in the penal nursery had more physical and social stimulation, owing to their full-time care by their own mothers. In the foundling home care was provided by one nurse, and infants were confined to their cribs, without human contact, for most of the day. Over the two years of study Spitz found that the foundling home children became progressively delayed in their physical and mental development compared with the nursery infants and a control group of home reared infants. At about age 3.0 years the foundling home children had average heights and weights expected for children aged 1.5 years. The developmental status of the nursery infants did not differ significantly from the controls. Moreover the mortality rate for infants in the foundling home was 37 per cent, while in the nursery group no child had died. Spitz used the term 'hospitalism' to describe the syndrome of poor physical and mental development and high mortality experienced by institutionalized children.

In a fascinating example of serendipity in scientific research, Widdowson (1951) found that the psychological well being of German children, orphaned during World War II, directly affected their growth. Children in two orphanages, 'Bienenhaus' and 'Vogelnest,' took part in a year-long nutrition supplementation experiment. Children of both orphanages were fed their usual ration for six months. In the second half of the study, the children of Vogelnest were given the usual ration plus additional, unlimited servings of bread, jam, and orange juice, but no supplement was given at Bienenhaus. Unexpectedly, during the first six months, when both groups received identical rations, the children of Vogelnest put on more weight, an average of 1.4 kg, than the children of Bienenhaus, an average less than 0.5 kg. Six months later the situation was reversed. Despite receiving extra food, the children of Vogelnest gained less weight (average gain of about 1.2 kg) than the children of Bienenhaus (average gain greater than 2.5 kg). However, eight of the children had gained an average of about 4.2 kg in the supplemented orphanage. Quite coincidentally, they had been transferred from Bienenhaus to Vogelnest just at the time when the supplement was given to Vogelnest. The records also revealed that the headmistress transferred between orphanages at the same time. Further inquiry by Widdowson found that the headmistress, a women referred to as Fraulein Schwarz, was a stern disciplinarian, who severely punished children for even minor infractions of behavior. She delivered these punishments at meal time, possibly to provide an example to the assembly of children gathered for the meal. The few children with favorable growth under the care of Fraulein Schwarz, were her 'favorites' who received no punishments, and when she transferred between orphanages she took her favorites along. Only these children, provided with positive psychological stimulation, showed the benefits of the extra food. Widdowson concluded that the emotional environment of the orphanages had a greater influence on physical growth than did the nutritional environment.

A serious limitation of the German orphanage study is that carefully controlled observations of behavior, food consumption, and energy expenditure did not take place. The results of the study cannot, therefore, be accepted at face value. Indeed, later studies challenged the idea that the emotional environment has a direct influence on growth. Sills (1978) reviewed the case histories of 185 patients hospitalized with the diagnosis 'failure to thrive,' a syndrome characterized by stature and weight below the third percentile for age, but without obvious organic cause. Detailed interviews with the parents, and home visitations, revealed that in more than half the cases parental neglect and deprivation of the child were the causes of the retarded growth. Withholding food, and outright starvation, were found in many of these cases of deprivation. Recent studies of adequately fed children with growth failure, living in emotionally disturbing environments, show that emotional stress can depress the secretion of digestive enzymes and the absorption of food, so that malnutrition may take place even if sufficient food is available (Parisi & de Martino, 1980). Thus, there is a possibility that many instances of psychological growth failure are actually cases of undernutrition caused only indirectly by a negative emotional environment.

Powell *et al.* (1967a, b) and Saenger *et al.* (1977) described a type of child with growth retardation simulating hypopituitary dwarfism, that is, short stature due to insufficient GH, that when removed from an emotionally stressful home environment has a total disappearance of symptoms. Usually, these children were admitted to hospital for the treatment of behavioral as well as physical symptoms. Before any treatment was administered, however, they experienced a spontaneous improvement in behavior, an increase in growth rate, and a rise in serum GH levels. Upon return to their home environments growth rate and GH levels both usually reverted to their previous low levels. Powell *et al.* (1967a) noted that malnutrition and stress-induced intestinal malabsorption were not primary factors accounting for growth retardation in these patients. The conclusion was that the low GH levels and retarded growth of these children are due more to emotional abuse than to physical neglect.

Another case study was reported by Magner *et al.* (1984) of a 12-year old boy who suffered growth retardation and delayed sexual maturation following an emotional trauma. The trauma was provoked by an argument between the boy and his stepfather, with whom the boy had a warm relationship. After the argument the boy verbalized a wish for his stepfather's death, and the next day the man seriously injured himself falling from a roof. The hospital where the man was recovering sent an erroneous 'notice of death' letter to the family's home which the boy received and read while at home alone. Though the man eventually recovered, the boy began a self-imposed period of food refusal and vomiting. He dropped from 34 kg to 25 kg in five months. At age 15 years, following periods of hospitalization and drug treatment, his eating behavior returned to normal, though at age 17 years he had the height of a normal child of 11.3 years, a bone age of 13.0 'years,' and was, essentially, prepubertal in physical appearance. He was given treatment with growth hormone at age 19.3 years, and between ages 20 and 21 years experienced a growth spurt and sexual maturation. Growth in height continued until age 25 years, when the young man reached 171.0 cm. The authors of this report state that in this patient, an acute 'psychic trauma induced a deranged hormonal state that persisted for several years' (p. 741). Though malnutrition and drug treatments in the three years following the trauma may have also upset the hormonal balance, the boy was behaviorally normal and drug-free for about five years before treatment with GH returned his growth and maturation to normal. This case, and the others previously discussed, exemplify the intimate and powerful relationship of emotional and hormonal factors in growth.

Green *et al.* (1984) reviewed the literature relating to psychosocial dwarfism, with an aim towards evaluating the role that nutrition and endocrine factors play in the etiology of the disease. It was found that most cases of growth failure in children under three years old were due to malnutrition, these infants usually being denied food by their emotionally disturbed mothers. Children over three years old were usually not clinically malnourished. Moreover, it was commonly found that both GH and IGF levels were significantly depressed in these older children. Since, as shown above, malnutrition is associated with low levels of serum IGFs and abnormally high secretion of GH, the endocrine profile of these children does not fit with starvation as the cause of their growth failure. Patton & Gardner (1963) proposed that emotional stress may affect some of the higher brain centers, particularly the amygdala and limbic cortex, which are known to control the emotions. Nerve impulses from these brain centers may pass to the hypothalamus, where they are transduced into neuroendocrine messages that may affect the production and release of hypothalamic hormones. In this manner, psychological disturbances in the child might be translated into a cutoff of GHRF in the hypothalamus, a halt in GH secretion from the pituitary, and depressed levels of IGF secretion from the body tissues. Green *et al.* favor this hypothesis, yet point out that it remains to be adequately tested.

7 Mathematical and biological models of human growth

something of the use and beauty of
mathematics I think I am able to understand.
D'Arcy Thompson, 1942

The treatment of human growth in Chapters 1–6 of this book emphasized the biology of development, for instance, anatomy, physiology, evolution, and ecology, with only brief excursions into mathematics, as was needed, to illustrate certain points of information. A scientific understanding of the pattern of human growth requires detailed information of the biological factors that determine development. However, it also requires the precision and economy of analysis that is provided by mathematics. These are not separate realms of knowledge, for, as may be seen in Figure 7.1, the growth and form of an organism may display, in quite an obvious way, a relationship between biological structure and mathematical regularity. D'Arcy Thompson (1942) described the biological form of the growth of the *Nautilus* shell, and several other spiral shapes in nature, with a mathematical function called an equiangular spiral. In a similar fashion, the form of the curve of growth of the human being can also be reduced to one or more mathematical functions. The mathematical treatment of human growth, or the growth of any other organism, is made possible by the predictability of biological development. Growth must produce a biological form that meets the ecological requirements of life for the species. Thus, in terms of growth and form, including the morphology and physiology of organisms, new individuals resemble other members of the same species more than they resemble members of other species. Owing to this predictability, growth and form are amenable to the precision of mathematical description.

Models in biology: description and explanation

Models are representations that display the pattern, mode of structure, or formation of an object or an organism. A model may also serve as a standard for comparison, between hypotheses that test human understanding of how some physical or biological phenomenon operates and the actual nature of that phenomenon. A complete understanding of

195

the regulation of human growth and maturation is made difficult by the many genetic, endocrine, and environmental factors, and their inter-actions, that influence the developmental process. Models are often employed to study aspects of human growth, such as the adolescent growth spurt, and aspects of maturation, such as the onset of puberty. Since these models represent only small portions of total development, they are easier to construct and easier to test than more comprehensive models.

Two types of models are used by growth researchers. The first type describes a result. For instance, a series of longitudinal measurements of the height of a child may be compared to a mathematical formula that fits a curve to the growth data. The mathematical parameters of the curve model the increases in height of the child over time. The fit of the curve may be quite precise; however, this type of model does not explain why increases in height occur or when changes in the rate of growth in height are likely to take place.

Figure 7.1. Shell of *Nautilus pompilius*, sagittal section (from D'Arcy Thompson, 1942).

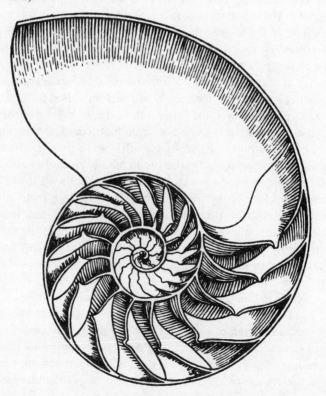

A second type of model attempts to describe a result as well as explain some of the determinants of the observations. An example is D'Arcy Thompson's model for the growth of the *Nautilus*. The equiangular spiral, the mathematical function that describes the growth of the *Nautilus* shell, implies that the proliferation of shell material occurs at a constant rate; proportional to the amount of tissue already produced. Thus, volumetric growth rate constantly accelerates, producing ever-larger chambers in the shell. Unlike a purely descriptive representation, D'Arcy Thompson's model predicts, with great precision, the size and volume of the next chamber to develop.

A predictive model of growth has both mathematical and biological meaning, and is preferred to descriptive models. Applied to human growth, such a model might attempt to describe and predict growth in height from birth to maturity. For instance, if it is assumed that a genetic program aims the growth of a child toward an 'ideal' size, as might be the case with strong genetic selection for size, then the rate of growth should be proportional to the difference between present size and ideal size. Departures from the predicted rate of growth, such as the mid-growth spurt during childhood and the adolescent growth spurt, may be explained by additional hypotheses about the regulation of the growth. In this way a model building process is encouraged, which leads to the testing of hypotheses against observations of growth and maturation and the eventual formulation of a theory of human development. In the discussion that follows, some of the major mathematical and biological models of human growth are reviewed and evaluated in terms of their contribution to growth theory.

Mathematical models of human growth

More than 200 mathematical formulae now exist that model some aspect of human growth and development (Timiras, 1972). Of these, about half a dozen are widely used for the analysis of human growth (Marubini, 1978). Count (1943) analyzed the growth in height of a sample of Chinese males and females between the ages of birth and 21 years. Using the distance curve for stature, Count fitted three mathematical functions to the empirical data. The first was a logarithmic curve ($y = a + bt + c \log t$, where y equals stature, a, b, and c are constants, and t equals time) which fitted the data from birth to about age seven years. The second was a simple step-up in the velocity of the logarithmic curve ($y = k$(value of logarithmic curve) $+ q$, where k and q are new constants) which fitted the data from about age seven to age 13 years. The third was a logistic curve ($y = v + 10 \exp[-pB] \exp[-r]$, where v, p, B, and r are constants and B includes a time parameter) which fitted the data from about age 13 years to the attainment of adult height. Count associated the three mathematical curves with biological periods of development. The first curve, and developmental period, corresponds to infancy and early childhood and this curve ends at about the time of the final formation and eruption of the first permanent molar tooth. The second curve, and period, corresponds to later childhood and this curve ends at about the time of the eruption of the second permanent molar. The third curve, and period, includes the adolescent growth spurt and the rapid decrease in the velocity of growth following the adolescent spurt (hence, the negative values for the exponents in the equation). This last curve ends at about the time of the eruption of the third molar.

Following Count's presentation of the three-curve model, other researchers proposed models of growth with more or fewer functions. Quo (1955) suggested a four-curve model, which was similar to Count's model except that infancy (birth to nine months) and early childhood (nine months to five years) were divided into separate growth periods. Quo argued that this model corresponds closely with endocrine events that underlie the pattern of growth in height. Deming (1957) proposed a two-curve model that, essentially, presented the process of growth in terms of a prepubertal component and an adolescent component. For the prepubertal component, from birth to nine years in girls and 10 years in boys, Deming suggested the logarithmic function. For the adolescent component, from age nine or 10 to maturity, a form of logistic function (often called the Gompertz (1825) curve) was suggested. This logistic function has the form $y = a \exp\left[-\exp\left(b - ct\right)\right]$. Deming showed that these two curves can represent accurately empirical data for stature from the prepubertal and adolescent periods of growth. Marubini *et al.* (1972) compared the goodness-of-fit of the logistic function used by Count and the Gompertz curve used by Deming to model adolescent growth. Although both functions produced a reasonably good fit, with small residual errors (the deviations between the fitted curve and the empirical data), the logistic function was found to have 'slightly better' statistical precision and robustness. Consequently, Marubini *et al.* recommended the use of the logistic function to estimate growth in height, and other skeletal dimensions, during adolescence.

Modeling the transition between growth periods

A major problem with each of the mathematical models of growth so far proposed is how to make the transition between the end of one period of growth and the beginning of the next. Count, Quo, and Deming suggested approximate chronological ages to mark the transition from one growth period to the next and, therefore, the time when the use of one growth curve should give way to another. The growth of individual children, however, more closely follows a biological timetable than a calendar schedule. Count's association of molar tooth eruption with the transitions between his three curves of growth is a step towards the use of biological criteria to mark the time when the switch between curves should occur. Similarly, Quo's discussion of endocrine changes from one period of growth to another is an attempt to use biological events to mark the transition. A problem still remaining with these models is that there is no satisfactory way to make a smooth transition between mathematical curves that have different functional forms.

Bock *et al.* (1973) proposed a new model of human postnatal growth to solve the problem of making a mathematical transition from the prepubertal period to the adolescent period of growth. The authors built on ideas proposed by Robertson (1908) and Burt (1937), that human growth could be described with the use of two or three logistic curves. Since these curves have the same functional form they may be added together to produce a smooth model of human growth. Bock *et al.* (1973) developed a double-logistic function to describe growth in length. This model is the summation of two logistic curves: the first describes 'a component of prepubertal growth which continues in reduced degree until maturity, and the second term describes the contribution of the adolescent spurt' (pp. 64–5). This model fits the distance curve of growth with reasonable precision, but does not estimate well velocities of growth in height. Bock & Thissen (1976) refined the model by adding a third logistic term which, essentially, divided the prepubertal period into two components. The addition of the third term improves the fit of the model to both the distance and velocity curves of growth. Unfortunately, the triple-logistic model has 10 parameters to be estimated, and adult height must be known. In many cases, the analysis of longitudinal growth data for children does not include a measurement of final adult height and, often, there are ten or fewer measurements of growth. Thus, more mathematical parameters must be estimated than there are empirical data points, which is statistically undesirable from both a practical and a theoretical standpoint.

When appropriate data are available, the triple-logistic model proves to be useful in describing distance and velocity curves of growth. The model is also of value since different growth periods, such as infancy, childhood, and adolescence, may be described with precision. Velocities of growth during each period, the age of maximum velocity, and the contribution of each period of growth to final adult height may be estimated. For example, the triple-logistic model was used to produce and analyze the growth curves depicted in Figures 6.1 and 6.2 of this book. Furthermore, the transition from one growth period to the next is smooth, and the contributions of each period may overlap, that is, the contribution of prepubertal growth is still active during the early phase of the adolescent growth period. Bock & Thissen (1980) speculate that the genetic and endocrine determinants of development that characterize each period of growth operate in a similar fashion; making a smooth transition, and overlapping from one period to the next.

A purely mathematical model of growth

Preece & Baines (1978) derived a series of mathematical functions that describe the distance and velocity curves of growth in height from the age of two years to maturity. The authors describe their method of derivation as, 'purely empirical and has made no pretense to true biological meaning' (p. 17). Marubini & Milani (1986) describe the Preece–Baines curves as being based on the assumption that the rate of growth is proportional to the difference between height at any age prior to maturity and height at maturity. The rate of growth is not a simple constant of proportionality, rather it is a function of age. This means that at different ages the rate of growth may be relatively slow, or relatively fast, compared with other ages, which is the manner in which children actually grow. A set of differential equations was used to calculate the age function for the rate of growth, and the solution yields three functions, of which the following is preferred by Preece & Baines for application to growth data:

$$h = h_1 - 2(h_1 - h_c)/\{\exp [s_0(t - c)] + \exp [s_1(t - c)]\}$$

where h is height at time t, h_1 is final (adult) height, s_0 and s_1 are rate constants, c is a time constant, and h_c is height at $t = c$.

Although the model was derived empirically, Preece & Baines were able to correlate each of the five parameters of the model with 'biological' events that occur during growth. The rate constants s_0 and s_1 correlate the strongest with the minimal prepubertal velocity of growth and the peak growth velocity during adolescence, respectively. Time c has a very high correlation with the age at peak height velocity during the adolescent growth spurt and h_c has a similarly strong association to height at peak height velocity. The relationship of these parameters to the velocity curve of growth is illustrated in Figure 7.2. By computation, other growth events may be estimated, such as age at the minimal prepubertal velocity (MPV), height at MPV, and amounts of growth between MPV, peak height velocity, and final height. The values of these estimates are useful when comparing the pattern of growth of one individual to another or of one population to another. In Figure 4.1 the velocity curves of growth in height for Australian, British, and African children were compared using the Preece–Baines function. Hauspie *et al.* (1980b) used the same function to estimate the value of several growth parameters for Belgian girls, and Hauspie *et al.* (1980b) used the function to describe the growth of a sample of children from India.

Each of the mathematical functions described above is parametric, that is, the form of the curve to be fitted to a series of growth measurements is defined by a set of predetermined variables, or parameters. For the Preece–Baines equation there are five parameters: h_1, h_c, c, s_0, and s_1. For the triple-logistic equation there are 10 parameters. A curve of growth is fitted to the measurements of height of an individual or a population by estimating the value of these parameters, most often by the use of non-linear regression. The parametric models are very useful for describing major growth events, such as the minimal prepubertal height velocity and the intensity, duration, and timing of the adolescent growth spurt. However, the form of these models is arbitrary and defined *a priori* to the analysis. As Marubini & Milani (1986) point out, growth data are

Figure 7.2. Height velocity curve of a boy with parameters of the Preece–Baines model indicated. MPV = minimal prepubertal growth velocity; PHV = peak height velocity during the adolescent spurt. Other terms as defined in text.

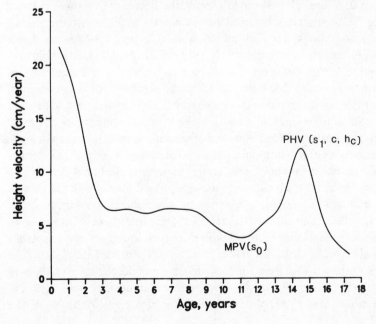

'forced' to conform to the shape of the model. This may obscure information about patterns of growth that do not conform to the constraints of the
chosen parameters. For example, of all the parametric models presented
above, only the triple-logistic function describes the mid-growth spurt in
height that occurs in the prepubertal growth of many children.

Non-parametric models of growth

One alternative to parametric growth modeling is through the
use of a non-parametric curve-fitting technique. This approach fits a curve
to a set of growth measurements for an individual without any prior
assumption as to the form the curve should take. After a satisfactory fit is
achieved, the value of growth events such as the duration, intensity, and
timing of the adolescent growth spurt may be calculated from the fitted
curves. Largo *et al.* (1978) used a smoothing spline function to fit the
longitudinal measurements of growth in height of 112 boys and 110 girls
measured in a Swiss study. A detailed description of the smoothing spline
algorithm and the statistical methods used to fit the growth data may be
found in the article by Largo *et al.* (1978). An example of a curve fitted to
growth data by smoothing splines was given in Figure 1.5. The curves
illustrated in this figure are very smooth, although they are fitted to only
about 35 data points ranging from the boy's birth to his 18th birthday. The
smoothing spline technique preserves the major features of the velocity
curves of growth. For instance, the adolescent growth spurt is easily
perceived, as are the mid-growth spurt, the decrease in growth velocity
before the adolescent spurt, and the asymmetry of the adolescent spurt.

The smoothing spline method has some drawbacks, one being that it
may significantly underestimate the value of the peak velocity of growth
during the adolescent spurt. This is because velocities and accelerations
in growth rate are calculated from linear segments of a curve, each
segment estimated in a piecewise fashion. Sharp changes in velocity and
acceleration are truncated, so as to make the line segments approximate
to a smooth curve. Owing to this and other limitations, Gasser *et al.* (1984;
1985) used another non-parametric technique for growth modeling,
namely kernel estimation. The kernel method is similar to smoothing
splines, but has the advantage of using weighted averages of the growth
measurements, in this case a series of longitudinal measurements of
height, rather than the raw values of the measurements. Gasser *et al.*
(1984) found that kernel estimates produce a smoother curve with smaller
residual errors (i.e., better fit) than the spline method. Kernel estimates
are also superior to splines for the calculation of acceleration changes in
the velocity curve of growth. The authors argue that velocities and
accelerations are closer to the underlying biological dynamics of the

growth process than are distances. For all these reasons, the kernel estimation method is preferred to smoothing splines.

Using kernel estimation, Gasser *et al.* (1985) analyzed the longitudinal patterns of growth in height for a sample of Swiss children (45 boys and 45 girls). The authors were interested especially in the relationship of the mid-growth spurt to the adolescent spurt, the structure of the adolescent spurt (its intensity, duration, and timing), and the role of the adolescent spurt in the determination of adult height. The mid-growth spurt and the adolescent spurt were found to be statistically independent in terms of time of onset, intensity, and duration. This finding provides mathematical support for the conclusion, based on endocrine studies and field research, that the mid-growth spurt and the adolescent spurt are independent biological events (see discussion in Chapter 6). The intensity, duration, and timing of the adolescent spurt were not correlated significantly with adult height. Rather, the velocity of growth during the prepubertal period was a significant predictor of adult height. The authors note that this finding is not unexpected, since children without gonadal function may reach normal adult height without experiencing an adolescent growth spurt (Prader, 1984).

The adolescent growth spurt represents a major quantitative and qualitative change from the pattern of growth that takes place during childhood. Since the intensity and duration of the spurt do not seem to be determinants of adult height or weight, Gasser *et al.* (1985) state that, 'it is an intriguing question what the role of the [adolescent spurt] is in the growth process' (pp. 512–13). They argue that the spurt may function to end growth. That is, the increased production and release of gonadal hormones that occurs during puberty at first increases the rate of growth. These hormones also increase the rate of epiphyseal closure of the long bones and, therefore, their long-term effect is to stop growth in height. The kernel estimate analysis of the longitudinal data showed that there is a correlation between the maximum acceleration of growth at the start of the spurt and the maximum deceleration of growth after peak height velocity. This means that children with more rapid growth during the initial phase of the spurt also have a more rapid termination of growth during the latter part of the spurt. In this way, Gasser *et al.* suggest, the adolescent spurt may allow a 'last chance' for some small amount of regulation of growth in height by variation in the duration of the spurt. For instance, a short child may achieve average adult height by having a less pronounced acceleration in growth velocity at the beginning of the spurt. Such a child will tend to have a longer lasting spurt, a longer lasting period of growth, and will grow taller than another short child who has a more intense adolescent spurt.

The notion that the adolescent spurt serves to end the growth period was also proposed by Count (1943). Count noticed that if the first of the three functions he used to describe growth in height, the one that he fitted to the curve of growth from birth to about seven years of age, were extended into adulthood it would reach the same final value for height as the three-function curve. However, there would be no adolescent spurt and adult height would be achieved at a significantly later age. Count concluded that the effect of the adolescent spurt is to bring the child to adult stature at a relatively early age, and bring about an end to growth in height.

An evolutionary and ecological interpretation of the growth spurt was presented in Chapter 3. There it was argued that the rapid increase in growth rate during adolescence, and the sequence of pubertal maturational events specific to each sex, lead to the appearance of reproductive maturity in girls before they are, in fact, fertile. For boys, fertility precedes the attainment of adult size and the development of secondary sexual characteristics. The sex-specific sequence of adolescent maturational events helps girls and boys to learn the economic, social, and psychological skills required for successful reproduction. Thus, in terms of biology and behavior the adolescent spurt is a major growth event with far reaching consequences. Indeed, the sexual, economic, and social implications of the adolescent spurt seem to be more important than the rather minor function it may have for 'fine-tuning' final height. Perhaps, in answer to the question of Gasser and his colleagues, the role of the adolescent growth spurt may be found in its relation to the biology and psychosocial maturity required for human reproduction.

The 'switch-off' model of growth

Another attempt at a non-parametric model of human growth was developed by Stutzle *et al.* (1980). The authors modeled the growth process from birth to adulthood with two mathematical components, one associated with childhood growth and the other associated with adolescent growth. Two ways were examined in which to link together the prepubertal and adolescent components of the growth process. The first is by an additive mathematical relationship, where the prepubertal component contributes also to the adolescent component. The second way is by a 'switch-off' model, where the onset of the adolescent component terminates the contribution of the prepubertal component. Stutzle *et al.* chose the 'switch-off' model because, they believe, it is more consistent with empirical observations of human growth in height than is an additive model.

The form of the 'switch-off' function is:

$$v(t) = a_1 S_1[(t - b_1)/c_1]F_1[(t - b_2)/c_2] + a_2 S_2[(t - b_2)/c_2]$$

The term $v(t)$ is the velocity of growth expressed as a function of age (t). The first term on the right hand side comprises a prepubertal curve shape function (S_1) and parameters for the intensity (a_1), time (b_1), and duration (c_1) of prepubertal growth. The last term on the right hand side, following the addition sign, is an adolescent curve shape function (S_2) and its three parameters for intensity, time, and duration of the adolescent growth spurt. F_1 is a link function between prepubertal and adolescent growth that smoothly joins the two components together. Coupled in location and scale to the adolescent component, the link function gradually decreases to zero the contribution of the prepubertal component.

Details of the choice of initial shape function (S_1 and S_2) and the link function are given in Stutzle *et al.* (1980). The best fitting curves for the entire growth period are estimated by iterating the shape functions against longitudinal growth data for individual children until a predetermined level of error between the empirical data and the fitted curve is satisfied. The prepubertal and adolescent curves are estimated separately, and later combined by the link function. The authors applied their model to data derived from a Swiss growth study. The sample included 45 boys and 45 girls measured from age one to 20 years. An example of the application of the 'switch-off' model is presented in Figure 7.3 for the growth of a hypothetical boy with the median values for the sample of the prepubertal and adolescent parameters. The solid line models the velocity curve of growth from age one to 20 years. The dotted line represents the termination of the prepubertal component after it is 'switched-off' by the onset of the adolescent component (the dashed line). Apparent in this example are the mid-growth spurt, which is entirely due to the prepubertal component, the dip in velocity prior to the onset of adolescence, and the asymmetry of the adolescent growth spurt. The negative velocity in the adolescent component is, according to the authors, an indication that the transition from prepuberty to adolescence is not adequately determined. This is due, possibly, to the 'choice of a rigid coupling between the two components via the [link] function' (p. 523).

The fit of the 'switch-off' model to the growth of individual children, both boys and girls, is similar to that depicted in Figure 7.3. The 'switch-off' model proves to be better than additive models in reflecting biological qualities of empirically derived growth curves. For instance, in the 'switch-off' model the prepubertal component of growth velocity is statistically identical in intensity, time, and duration for the median boy and girl. The parameters of the adolescent component differ significantly between the sexes. The estimated parameters of additive models, e.g., the triple-logistic of Bock & Thissen (1980) and the smoothing spline of Largo *et al.* (1978), differ significantly between boys and girls in both the prepubertal and adolescent components. In real life, as in the case of the 'switch-off' model, there is little difference in the velocity of growth in height during childhood between the average girl and the average boy. Another attractive feature of the 'switch-off' model is that the statistical correlations between the prepubertal and adolescent parameters of growth are not significant. There is a significant negative correlation, however, between the intensity and time parameters of the adolescent component. This indicates that early maturing children tend to have a larger value for peak height velocity, and late maturing children tend to have a smaller value. One would like to know why this should be so, but the model only describes these relationships and does not offer a theory of growth. However, these correlations conform closely to empirical growth data; prepubertal growth events have little relation to adolescent growth events, and early maturing children tend to have a more intense growth spurt than later maturing children (Simmons & Greulich, 1943; Tanner, 1962). In contrast, additive parametric models (Bock & Thissen, 1976) and non-parametric models (Largo *et al.*, 1978; Gasser *et al*, 1985) indicate, incorrectly, a statistical independence of the timing and intensity of the adolescent growth spurt.

Figure 7.3. Growth velocity curve for the height of a 'median' boy, based on the 'switch-off' model of Stutzle *et al.* (1980). See text for explanation (from Marubini & Milani, 1986).

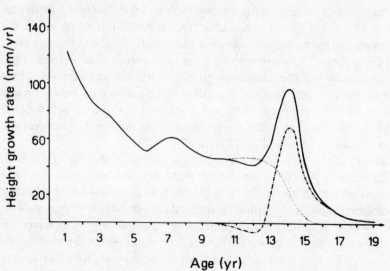

Finally, the 'switch-off' model was applied conceptually to the process of adolescent growth in both normal and pathological development. The link function (F_1) of the model smoothly terminates the contribution of the prepubertal component after the initiation of the adolescent component. Stutzle *et al.* believe that this is 'in good accordance' (p. 523) with the normal endocrine changes of puberty, especially the increased secretion of gonadal hormones, that bring about a closure of the epiphyses of the long bones and an end to growth in height. In the case of endocrine pathology, the link function also seems to reflect the growth of children empirically observed. Children with eunuchoidism, or prepubertal castration, do not experience an adolescent spurt in growth. However, the growth period is extended past the age of 20 years. These individuals reach a normal adult height, and apparently this is achieved entirely by the extension of the prepubertal component of growth, which is never 'switched-off' owing to the lack of puberty. In another pathological condition, children with precocious puberty (the onset of gonadal maturation prior to eight years of age) are taller than their age mates who are still prepubertal. As adults, however, these individuals are below average in height. The initial increase of height and later shortness associated with precocious puberty are explained by the 'switch-off' model as a function of the early onset of the adolescent component of growth, which temporarily elevates the height of affected children, but later depresses it owing to the early cut-off of the prepubertal component. With less time for prepubertal growth, adult height is significantly reduced.

Biological models of human growth

The 'switch-off' model of Stutzle *et al.* is more realistic biologically than most of the other models of growth so far discussed. The major fault of the model is the initially negative velocity of the adolescent component of growth. As mentioned above, this biological impossibility may be due to the constraints of the link function that forces the prepubertal and adolescent components to interact mathematically in a smooth and continuous fashion. By implication, it is suggested by Stutzle *et al.* that, biologically, prepubertal and adolescent growth also behave in a continuous fashion with adolescence smoothly inhibiting the contribution of the prepubertal period of growth.

If the mathematical form of the link function is not adequate to describe the onset of adolescence, perhaps the biological implications of the function are inadequate as well. Largo *et al.* (1978) noted that 'prepubertal growth is completely different from pubertal growth' (p. 422). Similar observations have been made by other students of human growth (Bertalanffy, 1960; Mellits & Cheek, 1968; Bock *et al.*, 1973; Frisch, 1974;

Habicht *et al.*, 1974; Bogin, 1978). Examples of human growth supporting this point of view were presented in Chapter 4, especially those cases of differences in the sensitivity of growth during childhood and adolescence to environmental and genetic determinants.

Growth in height, or any other physical dimension, is a continuous process and the parametric and non-parametric models of growth are able to describe this process mathematically with great precision. If the sole purpose of growth modeling is quantitative description, these models are sufficient. However, if the purpose of growth modeling is to develop hypotheses that are aimed at revealing the biological determinants of growth, then another approach must be taken. Assuming for the moment that the patterns of prepubertal and adolescent growth are 'completely different', it may be suggested that they are regulated by biological mechanisms that are independent of each other.

A conceptual model of growth

Tanner (1963) proposed a conceptual model for the biological regulation of human growth. Basic elements of the model are represented in Figure 7.4. A major feature of this model is that growth is target seeking and self-stabilizing. The curve labeled 'Time tally 1' represents a hypothetical mechanism that provides a 'target size' for body growth, and also keeps track of biological time during infancy and childhood. Biological time is measured in units of maturation, with the clock started at conception and stopped when some functionally mature state is reached. The curve labeled 'inhibitor' represents the concentration in the body of a hypothetical substance, perhaps a byproduct of cell division or protein synthesis, that acts upon the time tally to regulate growth rate. The amount of mismatch, M, between the two curves determines the rate of growth at each chronological age.

Tanner's model accounts for the deceleration of growth velocity during infancy and childhood and explains the phenomenon of catch-up growth (Prader *et al.*, 1963) following serious illness or starvation. During the normal postnatal growth of an infant, the amount of mismatch between the size the infant might attain and its actual size is large, and growth rate is rapid. As the child grows in size the amount of mismatch decreases and the concentration of inhibitor increases. As a result, the rate of growth slows. Under non-normal conditions, for instance, starvation, the rate of growth slows or stops during the period of insult. The concentration of inhibitor remains constant during this time as well. The time tally continues to register the mismatch between actual size and target size. When the insult to the child is removed, in this case upon refeeding, there is a rapid increase in growth rate to restore the balance between the time tally, the expected amount of mismatch, and the concentration of inhibitor. When this balance is restored the rate of growth assumes the normal velocity for that child, as if the insult had not occurred.

To account for the abrupt change in the velocity of growth that occurs at adolescence, Tanner suggests the existence of a second time tally. 'Time tally 2' (Figure 7.4) operates in the same manner as time tally 1, but both the mechanism controlling the new tally and its inhibiting substance are assumed to be distinct from the old one. Tanner believes that the switch to the new tally occurs when a minimum velocity of growth, or minimum mismatch, on the old tally is reached. This is labeled point *L* in Figure 7.4. After the switch occurs a new larger mismatch is established and a rapid increase in growth rate results. This is the adolescent growth spurt. As the mismatch is reduced, and the concentration of inhibitor is increased, the rate of growth slows once more. Variation in the timing of the adolescent spurt between individuals is explained by changing the point at which the switch between tallies takes place.

This model is as conceptually stimulating today as when it was first proposed. However, scant attention has been paid to it and little research

Figure 7.4. Tanner's conceptual model for the regulation of human growth. Rate of growth is determined by the amount of mismatch (*M*) between the concentration of a hypothetical inhibitor substance and a time tally. Time tally 1 controls growth during childhood. Time tally 2 controls growth during adolescence. At point *L* the switch between time tallies occurs, and the adolescent growth spurt is initiated (from Tanner 1963).

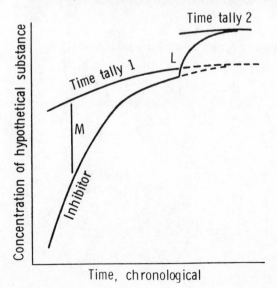

has been directed towards the description of the time tallies and the inhibitor substance. The reasons for this may be that the model conflicts, in at least four ways, with the more widely used parametric and non-parametric growth functions. Tanner's model conflicts with these in that (1) it is explanatory and predictive rather than only descriptive, (2) its parameters, though speculative, are biologically meaningful, (3) it is entirely qualitative rather than quantitative, and (4) it proposes a discontinuous relationship between the factors that regulate growth before and after puberty. The first two differences are desirable in any biological model of human growth. Such a model should be able to predict growth events and offer biological reasons for the predictions. The qualitative nature of Tanner's concept of growth and the discontinuous change it suggests are at odds with the conventional method of developing mathematical models for biological systems. This method uses the calculus, and by solving a series of differential equations it assumes a quantitative analysis and a system that behaves in a smooth and continuous fashion.

A catastrophe theory model of human growth

Another branch of mathematics offers a method of analysis whereby qualitative and discontinuous phenomena may be described. This method is derived from topology and is called catastrophe theory

Figure 7.5. A simple catastrophe, illustrated as a sudden change in potential energy (from Woodcock & Davis, 1978).

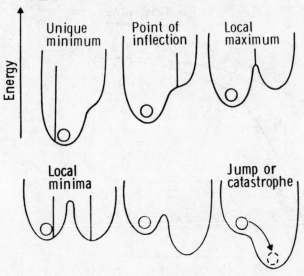

(Thom, 1975; 1983). Although the name implies disaster, the purpose of the method is to analyze phenomena in which a continuous change in the controlling variables results in a discontinuous change in their effect. Catastrophes, in the mathematical sense, include the change which comes from throwing an electrical switch, the breaking of a wave as it nears the shoreline, and the phase change that occurs when 'dry ice' (the solid form of carbon dioxide) is converted directly into a gas. In a physical sense, these catastrophes represent a continuous change in a set of controlling variables that eventually produces a sudden transition from one stable equilibrium to another. Figure 7.5 depicts this type of change between equilibria for a two-dimensional system with one controlling factor. Figure 7.6 illustrates a simple catastrophe in three-dimensional perception. The small circle is either in the center of one face of the cube or in the corner of another face. The transition is instantaneous and discontinuous with no sense of motion perceived from one stable configuration to the other. The neurological basis for this illusion is not known, but this example illustrates that discontinuous change is part of nature.

This 'either–or' type of catastrophe is the most simple and is formally called the fold catastrophe (Zeeman, 1977). There are, of course, more complex catastrophes, but Thom (1975) proved mathematically that for processes controlled by no more than four factors there are just seven

Figure 7.6. The Necker cube: an illustration of a simple catastrophe in perception.

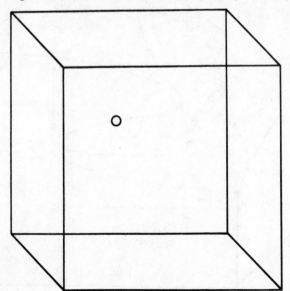

elementary catastrophes. Each of these has a unique shape and set of properties. One of these, the cusp catastrophe, is defined by two control factors and has a shape and set of properties similar to several growth phenomena.

Bogin (1980) explored some of the applications of the cusp catastrophe to human growth. An example of the cusp catastrophe is illustrated in Figure 7.7. It is a graph of a curved surface with a pleat, called the behavior surface, above a planar surface, called the control surface. Movement of the two control factors along the lines indicated in the figure defines a set of vectors on the control surface and a set of points, corresponding to these vectors, on the behavior surface. The points on the behavior surface represent pathways for smooth movement and discontinuous movement.

Mathematically, the cusp catastrophe is associated with the function

$$y = 1/4(x^4) - ax - 1/2(bx^2)$$

where a and b are the control factors and x is the variable whose behavior is plotted on the behavior surface. The topological features of the

Figure 7.7. An illustration of the cusp catastrophe model with pathways for continuous and discontinuous change (from Bogin, 1980).

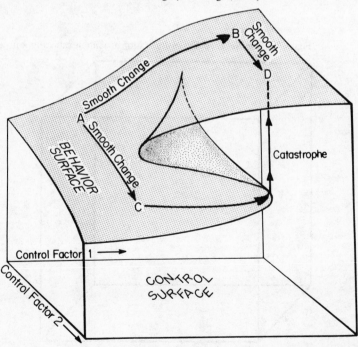

behavior surface represent 'the graph of all points where the first derivative of the function is equal to zero' (Zeeman, 1976, p. 78). The first derivative of the cusp function is $y = x_3 - a - bx$. Rigorous, but readable, mathematical discussions of catastrophe theory may be found in Zeeman (1976; 1977) and Woodcock & Davis (1978). The history of development of the theory and its metaphysical foundations may be found in Thom (1983). The glimpse of the mathematical foundations of catastrophe theory given here is meant to demonstrate that the graphic representations of the cusp catastrophe presented in the present discussion are predicted by the theory, and are not arbitrary configurations or post-hoc constructions.

When plotting the behavior surface, most combinations of control factor 1 and 2 result in a unique solution for setting the first derivative equal to zero. These unique solutions are points of stable equilibrium, or the most likely mode of behavior for the particular combination of control factors. The set of these points represent the areas that define the stable, non-pleated, parts of the behavior surface. For some combinations of control factors, however, there are two possible solutions for setting the first derivative equal to zero. Thus there are two stable equilibria and two likely modes of behavior. This is represented in the graph of the behavior surface as two overlapping sheets connected by a third sheet that together make a continuous pleated surface. The third sheet represents unstable equilibria in behavior, and its surface is generally inaccessible. Variation in the control factors in the region of the pleat will shift the behavioral variable between the lower and upper stable surfaces. Even though changes in the control variables are continuous and smooth, as reflected in the smooth continuity of the pleated surface, small changes in their relative levels may cause a sudden, discontinuous change in behavior. The discontinuous jump between stable surfaces is a catastrophe, in the mathematical sense. Both smooth and catastrophic change can occur with the cusp catastrophe model. Smooth transition between the upper and lower surfaces can take place, as in the movement between points A and B in Figure 7.7. Smooth transitions along the lower and upper surfaces may also occur, as in the paths from A to C and from B to D. Discontinuous change occurs when a path like that from point C to D is followed. Movement from D back to C will also result in a catastrophic fall to the lower surface when the edge of the upper surface is reached.

Tanner's conceptual model for the regulation of human growth is comprised of two time tallies, one for prepubertal growth and one for adolescent growth. The curves representing the time tallies may be thought of as line segments drawn from the lower and upper behavior surfaces of the cusp catastrophe model. In Figure 7.4 these lines are not connected. However, if this graph, which is two dimensional, is considered to be a 'slice' cut through the pleated region of the cusp model, it may be visualized how a third continuous line may be drawn to connect the two time tally curves (Figure 7.8). By expansion to three dimensions, a full visualization is achieved of the cusp catastrophe model as applied to human growth (Figure 7.9).

The factors controlling growth are numerous. However, each of the factors discussed in previous chapters of this book may be categorized roughly into those that control amount of growth and those that control rate of maturation. These two categories are chosen as control factors for the cusp catastrophe model of human growth. In Figure 7.9, the lower and upper behavior surfaces represent the pre-adolescent and adolescent patterns of growth and maturation, respectively. The shape of the behavior surfaces of the cusp catastrophe model reflects the shape of the velocity curve for human growth during childhood and adolescence. The rate of pre-adolescent growth declines, generally, until adolescence (the cusp catastrophe does not model the mid-growth spurt). Then there is a rapid increase in the rate of growth, peak growth velocity is achieved, and, finally, the rate of growth declines. The curve, or trajectory, labeled A in the figure parallels this sequence of changes in the rate of growth. The rapid increase in growth rate during the adolescent spurt is labeled as the 'Puberty catastrophe.'

Figure 7.8. Tanner's conceptual model converted into a two-dimensional cusp catastrophe by the addition of a pleated curve (from Bogin, 1980).

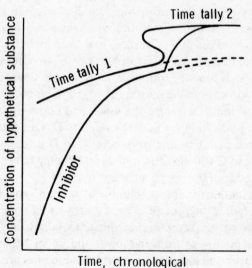

continuous stresses will show both a reduced intensity of the adolescent growth spurt and an increased duration of adolescent growth.

The representation of growth in three dimensions and the other topological features of the cusp catastrophe model provide a qualitatively satisfying description of human growth. It is possible to infer from the model some possibilities about the underlying biological mechanisms that actually regulate growth. For instance, in the region of the pleat the cusp catastrophe illustrates an overlap between the pre-adolescent and adolescent behavior surfaces. This suggests that factors responsible for growth patterns are operating simultaneously for some period of time. The smooth fold curve connecting the two surfaces implies that the variables regulating the change between pre-adolescent and adolescent patterns of growth act in a continuous fashion, even though the switch from one pattern to the next must be discontinuous.

The quantitative models of growth suggest that either a continuous transition occurs or a discontinuous switch between childhood and adolescence takes place. The triple-logistic model (Bock & Thissen, 1976) posits an additive overlap of the prepubertal and adolescent components of growth. Unfortunately, the constraints of this model force growth data to conform to a mathematical function rather than to biological mechanisms. On the other hand, non-parametric models do not impose these mathematical constraints and may represent more accurately the biology of growth. For example, the 'switch-off' model of Stutzle *et al.* (1980) has good concordance with many biological events that occur during growth. However, a weakness of the model is that it proposes that the contribution to growth of the prepubertal growth component ends, owing to the onset of the adolescent component. The link function that is used to join the two independently acting components does not adequately represent the velocity of growth during the transition (the adolescent component begins with a negative velocity).

The catastrophe theory model suggests that an overlap and a discontinuous switch between the mechanisms regulating growth take place. Using Tanner's metaphor of the 'Time tally' it may be imagined that the pre-adolescent behavior surface of the cusp model is controlled, in part, by the rate of maturation of Time tally 1. The adolescent behavior surface is controlled, in part, by the rate of maturation of Time tally 2. Tanner suggests that the time tallies are separate entities and that they are regulated independently. Thus, the onset of Time tally 2 at puberty may occur while Time tally 1 is still operating. A child's rate of growth may be influenced by both tallies, but the onset of Time tally 2 and the termination of Time tally 1 are sudden, discontinuous events. This concept of growth regulation at the time of puberty is consistent with empirical observations. As discussed in Chapter 4, and as shown above, there is much variability in the timing, intensity, and duration of adolescence between individual children and between populations. The three-dimensional shape of the catastrophe theory model allows for such variation in development, and the overlap and discontinuity of the mechanisms regulating rate of maturation may account for some of the causes of this variability.

Theories of growth regulation

The actual biological mechanisms involved in the regulation of amount of growth and rate of maturation are unknown. Tanner's model proposes that the control of amount of growth is related to the concentration of a hypothetical inhibiting substance that acts on a time tally located in the brain, perhaps in the hypothalamus. Goss (1978) suggests that overall size of the body is regulated by a genetically programmed amount of growth for certain visceral organs. For instance, the heart or the kidneys may grow to a predetermined size and the functional limits of these organs in supporting the operation of other organs and systems of the body may determine the size to which these tissues, and the body as a whole, may grow. Snow (1986) interprets the results of experimental studies of growth controlling mechanisms to support Goss' hypothesis. However, Snow finds that growth control in a developing organism is evident before organogenesis occurs during embryonic development. This fact, Snow suggests, indicates that control may lie in a tissue that differentiates relatively early in development and is distributed widely throughout the body, rather than in the visceral organs. A tissue composed of cells derived from the neural crest of the embryo, such as the central nervous system, is Snow's choice of a likely tissue for this function.

Rate of maturation is likely to be regulated by neuroendocrine factors (e.g., the interaction of the central nervous system and the hormonal system) of the body. Grumbach *et al.* (1974) and Grumbach (1978) reviewed the development of one neuroendocrine system, the hypothalamic–pituitary–gonadal axis, that occurs during human maturation. A possible chronology of the major neuroendocrine changes in this axis is outlined in Table 7.1. The events listed form a continuous series of changes between growth periods, but result in several discontinuities in growth and development, such as the adolescent growth spurt and menarche. This makes the neuroendocrinology of growth amenable to modeling by catastrophe theory. For 'whenever a continuously changing force has an abruptly changing effect, the process must be described by a catastrophe' (Zeeman, 1976, p. 80).

Table 7.1. *Postulated ontogeny of neuroendocrine activity of the hypothalamic–pituitary–gonadal axis*[1,2]

Fetal period
1. Maturation of gonadal hormone negative feedback mechanism after 150 days gestation.
2. Low level of GnRH secretion at term.

Infancy and childhood period
3. Negative feedback control of FSH and LH secretion becomes highly sensitive to low levels of gonadal steroids by two to four years of age.

Late pre-adolescent period
4. Decreasing sensitivity of hypothalamus to gonadal hormones.
5. Increased secretion of GnRH, independent of gonadal control.
6. Increased pituitary responsiveness to GnRH.
7. Increased responsiveness of gonads to FSH and LH.
8. Increased secretion of gonadal hormones.

Adolescent period
9. Continued decrease in sensitivity of negative feedback mechanism to gonadal hormones.
10. Sleep-associated increase in pulsatile secretion of LH and T.
11. Maturation of positive feedback mechanism; gonadal hormones stimulate secretion of GnRH and LH.
12. Spermatogenesis in males; ovulation in females.

---

[1] Modified from Grumbach (1978).
[2] GnRH, gonadotropin releasing hormone; FSH, follicle stimulating hormone; LH, luteinizing hormone; T, testosterone.

A catastrophe theory model of neuroendocrine maturation

Bogin (1980) modeled the chronology of neuroendocrine events with the cusp catastrophe. The model is illustrated in Figure 7.10. The point of origin of the control axes may correspond to a state of maturation that occurs about 150 days after fertilization. At that time the fetal hypothalamus becomes sensitive to the levels of circulating gonadal hormones. This sensitivity matures into a negative feedback system regulating the secretion of gonadotropin releasing hormone (GnRH). Low GnRH secretion diminishes the pituitary secretion of follicle stimulating hormone (FSH) and luteinizing hormone (LH), and this maintains gonadal hormone secretion at low levels. GnRH secretion is relatively low at birth, and the entire negative feedback system becomes highly sensitive and stable by two to four years of age. This system of neuroendocrine activity may, in part, regulate the pre-adolescent pattern of growth. As this system matures, the rate of growth during infancy rapidly diminishes, reaching a stable level between two and four years of age, which is about the same time as the negative feedback system stabilizes. During the late pre-adolescent period two changes occur. The first is the decreasing sensitivity of the hypothalamus to gonadal hormone levels in the circulation. The second is the initiation of a new GnRH secretion regulating system. There is evidence that the new GnRH system may be influenced by a changing hypothalamic sensitivity to morphine-like brain peptides, called the opiate peptides (Blank *et al.*, 1979).

These events are similar, in several ways, to Tanner's time tally model of growth. The new GnRH system can begin operating independently of gonadal function and the negative feedback control that is characteristic of the pre-adolescent growth period. For instance, clinical studies show that the late prepubertal increase in the secretion of GnRH, LH, and FSH (but not gonadal steroids) may occur in children with gonadal dysgenesis (the absence of functional gonads) at about the same time as in normal children (Boyar *et al.*, 1973; Grumbach, 1978). Because this increase in gonadotropic hormones is independent of pre-existing gonadal function, and since the gonadal hormones and the opiate-like peptides of the brain

Figure 7.10. Cusp catastrophe model of neuroendocrine ontogeny of the hypothalamic–pituitary–gonadal axis in relation to the control factors regulating human growth. GnRH, gonadotrophin releasing hormone; LH, lueteinizing hormone; T, testosterone (from Bogin, 1980).

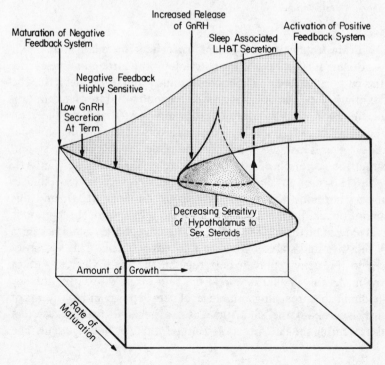

originate from widely divergent tissues in the body, it has been suggested that these neuroendocrine changes may provide a biological index for the initiation of puberty (Boyar *et al.*, 1972; 1973; 1974; Grumbach, 1978; Blank *et al.*, 1979). The pre-adolescent negative feedback regulation of GnRH may be likened to Time tally 1, while the change to the new pattern of GnRH regulation may be compared with Time tally 2.

An interaction between the pre-adolescent and adolescent systems of hypothalamic control may continue for some time, and this is reflected by the overlapping surface of the pleat in the catastrophe theory model. However, the influence of the adolescent system is marked by the rapid rise in the circulating levels of the gonadal hormones and in the development of primary and secondary sexual characteristics. The influence of the pre-adolescent system may be seen in the pattern of growth, which in some children continues along its pre-adolescent pathway even as some sexual development is taking place, until the sudden switch to the adolescent system occurs. Numerous studies, summarized by Marshall & Tanner (1986), show that there is much variation between individual children in the timing and the sequence of pubertal events, such as appearance of pubic hair, breast development, and the onset of the growth spurt in girls. Perhaps the clearest example of the overlap between the pre-adolescent and adolescent patterns is the difference in the typical sequence of adolescent growth and maturation events experienced by boys and girls. In Chapter 3 of this book it was noted that girls reach menarche after the peak height velocity of the adolescent growth spurt, but boys, usually, reach reproductive maturity before the peak height velocity of the spurt. This example shows, once again, that rate of maturation and amount of growth are independently controlled during development, and that each may be influenced separately by either the pre-adolescent or adolescent systems of regulation.

The work of Boyar *et al.* (1972; 1974) suggests that for boys, the switch between the two patterns of growth occurs when the adolescent system matures to the point defined by the sleep-associated increase in LH and testosterone (T) secretion. Presumably, girls may experience the switch at a similar point, only in this case the increase in LH secretion during sleep would stimulate the secretion of estradiol and other estrogens. Perhaps it is the difference in the serum concentration of estrogens and androgens, and tissue sensitivity to them, that leads to the difference in the typical sequence of pubertal growth events of girls and boys. In any event, it is at this point that the negative feedback control of hypothalamic GnRH secretion terminates and the influence of the pre-adolescent system ends. The growth trajectory cannot go back in time, or size, to pick up where the adolescent system began to operate, so the only alternative is a jump between the pre-adolescent and adolescent behavior surfaces. The adolescent growth spurt may be one of the observable results of the catastrophic change in the neuroendocrine systems regulating maturation and growth. As the adolescent system matures a positive feedback mechanism develops between the secretion of gonadal hormones and the stimulation of GnRH. In young men, GnRH secretion is maintained at a relatively high level from day to day, and this results in continual spermatogenesis. In young women, GnRH secretion occurs in a cyclical and pulsatile fashion, resulting in periodic ovulation. In terms of growth, rising gonadal hormone levels at first induce the adolescent growth spurt, but, eventually, bring about changes in the growth plate region of the long bones that cause growth in length to come to an end.

Since the catastrophe theory model was proposed in 1980, new research has resulted in a better understanding of the ontogeny of development of the hypothalamic–pituitary–gonadal axis. Kelch *et al.* (1985) and Marshall & Kelch (1986) summarized two new findings of their own research, and that of others, that modify some aspects of the catastrophe model. First, the action of opiate peptides of the brain has not been shown to inhibit the secretion of GnRH during childhood. However, the role of the opiate peptides in the stimulation of GnRH activity, and the onset of puberty, remains to be determined.

Second, the secretion of LH, and probably GnRH as well, in prepubertal children is not as different from the adolescent and adult pattern of secretion as previously believed. Children have both the pulsatile secretion of LH and the sleep-associated increase of GnRH and gonadotropin secretion. In children, however, the levels of GnRH secretion are very low, relative to the adolescent and adult levels. Thus, the onset of puberty is marked by an augmentation, and perhaps by an increased frequency of release pulses of GnRH 'rather than the *de novo* onset of GnRH release' (Kelch *et al.*, 1985, p. 113).

Advantages of a qualitative approach to the biology of growth

These newer findings may be used to modify some of the details of the catastrophe theory model. However, the application of the model to the neuroendocrinology of reproductive maturation is basically unaltered. The newer data still indicate that there is an overlap between the childhood and adolescent patterns of hormonal secretion. Moreover, puberty may still be viewed as a sudden and discontinuous shift between the negative feedback control of GnRH secretion that characterizes childhood and the positive feedback control of GnRH secretion that is present in adolescence and adulthood.

Many of the physical changes in the amount and rate of growth, of normal children and children with growth pathologies, are modeled by the cusp catastrophe. The same mathematical form may be used to model some neuroendocrine changes in development and are associated with growth and maturation. The coincidence of the two models shows, in a novel way, how growth and the physiological mechanisms that regulate growth may be related. This is the major advantage of the catastrophe theory approach. Quantitative formulae describe curves of growth without any biological meaning given, necessarily, to the parameters of the mathematical functions. These functions also assume a smooth continuity of the growth curves and of the mechanisms that produce the curves. The qualitative approach of catastrophe theory begins with an idea of how a system operates; in the present case the biological system of the growing child. Biological meaning is assumed at the start of the modeling process, rather than being constructed after an acceptable mathematical fit to growth data is made. As an example of the usefulness of the qualitative approach consider that Kelch and his colleagues believe that the current state of knowledge leads to the conclusion that the control of the onset of puberty resides in the operation of the central nervous system, including the hypothalamus, rather than in the pituitary or other endocrine organs. This was the view suggested by Tanner, based on his conceptual model of growth regulation. It is also the view reinforced by the catastrophe theory model, which combines a conceptual view of growth regulation with empirical data on the operation of the neuroendocrine system.

The qualitative approach has its own limitations. For instance, not all aspects of growth are accounted for by the catastrophe theory model. One of these is the mid-growth spurt. Of course, this is an event that is largely independent of growth at puberty, and also is independent of final growth in size. Thus, it may require a separate set of qualitative assumptions to model the mid-growth spurt. Another problem with the application of the catastrophe theory model is that the control variables, amount of growth and rate of maturation, are wholly qualitative. The model cannot be applied to the growth of individual children or populations until the control variables are quantified. A new branch of mathematics, known as discrete mathematics (Ralston, 1986), may be applicable to this problem. Discrete mathematics is quantitatively precise (it is the mathematics of computer science and graph theory). Unlike the calculus used so far to describe growth curves, discrete mathematics can provide numerical solutions to non-continuous functions. Perhaps an application of discrete mathematics to human growth, combined with the insights provided by catastrophe theory, may lead to a model of growth that is quantitatively and biologically valid.

Glossary

adolescent sterility A physiologic state of pubertal girls that begins after menarche and lasts until ovulatory cycles are established.

adrenarche The onset of secretion of androgen hormones from the adrenal gland, usually occurring at about six to eight years of age in most children.

age-graded play group A social group of children and juveniles in which the older individuals provide basic caretaker behavior and enculturation for the younger individuals. The play group frees adults for subsistence activities and other adult behaviors.

allometry Differential rates of growth of parts of the body relative to the growth of the body as a whole.

anthropometry The scientific measurement of the human body.

attachment behavior The set of physical and psychological cues and responses that bond an infant with one of its caretakers.

Australopithecus Genus of early hominids from eastern and southern Africa, dating to between four to one million years before present.

autocrine A hormone-like substance that has its biological effect on the cells that produce the substance.

auxology The study of biological growth.

biacromial breadth The linear distance between the shoulders, measured between the most lateral points of the acromial processes of the scapula.

bitrochanteric breadth The linear distance between the most lateral extensions of the greater trochanters of the femurs.

body composition The make-up of the body in terms of the absolute and relative amounts of adipose tissue, muscle mass, skeletal mass, internal organs, and other tissues.

calcification The process of mineral deposition, usually calcium, in tissues of the body. Regular sites of calcification are the skeleton and the dentition.

catastrophe theory A branch of mathematics that analyzes the nature of sudden and discontinuous change in a phenomenon that is produced by regular and continuous change in the factors that control the behavior of that phenomenon.

catch-up growth The rapid increase in growth velocity following recovery from disease or refeeding after short-term starvation.

childhood growth period A stage in the growth of human children that occurs between the end of infancy and the start of the juvenile growth period (about the ages two to ten years). Childhood is characterized by relatively rapid neurological development and slow physical growth and development.

cross-sectional growth study Measurement on a single occasion of individuals grouped by age and sex, and sometimes other characteristics.

224

development Progression of changes from undifferentiated or immature state to a highly organized, specialized or mature state.

distance curve A graphic representation of the amount of growth achieved by an individual over time.

dizygotic twins Twins that result from two independent fertilizations. Such twins are no more alike genetically than ordinary brothers and sisters. Also referred to as 'fraternal twins.'

dual effector model A hypothesis for the regulation of cellular growth that posits that the role of growth hormone is to cause the differentiation of fibroblast cells into specific tissue types (bone, adipose, etc.) and that the role of insulin-like growth factors is to cause the multiplication (clonal expansion) of the newly differentiated cells.

embryo Stage of prenatal development lasting from the second to tenth week following fertilization, characterized by the rapid differentiation of tissues and the formation of organs.

endocrinology The study of hormones, their origins and actions.

epidemiology The study of the causes and transmission of disease.

epiphysis An ossification center of a long bone, separated from the shaft of the bone by cartilage.

essential nutrient A nutrient that cannot be manufactured by the human body from simpler elements and thus must be supplied from the diet.

feedback control A method used for the regulation of biological activity in an organism based on the flow of information between parts of the organism. An example is negative feedback control in the endocrine system, in which rising levels of a hormone in the blood stream result in a lowering of the level of stimulation of cells that produce that hormone.

fetus Stage of prenatal development lasting from the tenth week following fertilization to birth.

fibroblast cell A type of undifferentiated cell which may differentiate into bone, adipose, or other types of tissue.

genotype The genetic constitution of an individual.

gonadarche Maturation of the gonads (testes or ovary) resulting in the secretion of gonadal hormones (androgens or estrogens).

growth Quantitative increase in size or mass.

heritability An estimate of the relative genetic contribution to the phenotypic expression of a physical or behavioral characteristic. The value of the heritability estimate is a function of genetic factors, environmental factors, and the interaction of genetic factors and environmental factors.

hominid Living human beings and their extinct fossil ancestors. Primate species characterized by habitual bipedal locomotion.

hominoid The group of the Primates that includes human beings, the apes, and their extinct ancestors.

homoiothermy Self-regulation of a relatively constant body temperature.

hormone Chemical substance secreted from a specific tissue into the general blood circulation, where it travels to its site of action.

hyperplasia Cellular growth by cell division (mitosis).

hypertrophy Cellular growth by an increase of material within each cell.

hypoxia The lack of sufficient oxygen supplied to the tissues of the body. May be the result of disease or may be due to residence at high altitude (3000 meters or more above sea level).

juvenile growth period A stage in the development of many higher primates, defined as the time prior to sexual maturation when an individual is no longer dependent on adults for survival.

karyotype A description of the chromosomes of an organism in terms of their number, shape, and size.

kwashiorkor A disease of protein deficiency, characterized by failure to grow, wasting of the muscles, loss of appetite, irritability, changes in the hair and skin, and anemia.

longitudinal growth study Measurement of the same individual or group of individuals, repeated at regular intervals.

low birth weight A weight at birth of 2500 grams or less for a neonate of normal gestation length.

marasmus A disease of severe undernutrition, especially energy and protein malnutrition, characterized by failure to grow, wasting of muscles, edema, apathy, and a ravenous appetite.

menarche The first menstrual period.

mid-growth spurt A relatively small increase in the rate of growth in height that occurs in many children between the ages of six and eight years.

migration The movement of people from place to place. In recent years, much of this migration has been from rural to urban areas, requiring substantial changes in the environment and lifestyle of the migrants.

mitosis Cell division resulting in two 'daughter' cells with the same genetic constitution as the original 'parent' cell.

model A representation that displays the pattern, mode of structure, or formation of an object or organism.

monochorionic placenta A type of placenta usually associated with twin pregnancies in which a single chorion (one of the placental tissues) is shared by both twins.

monozygotic twins Twins that result from a single fertilization and, therefore, share the same genotype. Also referred to as 'identical twins.'

neocortex Region of the mammalian brain associated with 'higher' level motor and sensory activities and the integration of these activities into complex patterns of behavior.

neonate Stage of postnatal development lasting from birth to 28 days after birth.

neoteny The retention of infantile or juvenile traits into adulthood. This is achieved by having sexual maturation take place while the individual is still in a pre-adult stage of phenotypic development.

organogenesis The formation of body organs and systems during the first trimester of prenatal life.

ossification The process of bone formation in skeletal tissue.

paedomorphism Having features in the adult like those of the child. Unlike neoteny, paedomorphism results in only the appearance of child-like features, which are, in fact, unlike the childhood condition in actual form and function.

paleoecology The study of extinct forms of life and their relations with the environment (e.g., types of foods eaten, requirements for reproduction).

paleontology The study of extinct forms of life, usually as represented by fossilized remains.

paracrine A hormone-like substance that has its biological effect on nearby cells within the same tissue as the cells that produce the substance.

parental investment The allocation of resources, such as time or energy, to offspring that occurs at some cost to the parents.

peak growth velocity The maximum rate of growth in height, weight, etc. achieved during the adolescent growth spurt.

percentile of growth Method of ranking growth status for height, weight, etc. of an individual relative to other members of a sample or population of people. Example: a child at the 75th percentile for height is taller than 75 per cent of the other children in the group under consideration.

phenotype The physical or behavioral appearance of an individual, resulting from the interaction of the genotype and the environment during growth and development.

placenta An organ of some mammals composed of fetal and maternal tissues that transfers nutrients and oxygen from the mother's blood circulation to the fetus and fetal wastes to the mother's blood circulation for disposal.

plasticity The concept that the development of the phenotype is responsive to variations in the quality and quantity of environmental factors required for life. Such variations produce many of the differences in growth observed between individuals or groups of people.

preformation Erroneous notion that the prenatal human body is essentially adult-like in form.

premature birth A birth that occurs prior to 37 weeks of gestation.

psychosocial dwarfism A type of growth retardation produced by a negative physical and emotional environment for growth.

secondary sexual characteristics Physical traits associated with the onset of sexual maturation, including the development of facial hair and muscularity in boys and the development of the breast and adult fat distribution in girls.

sedente A person living in his or her geographic region of birth; a non-migrating individual.

'seed eater' hypothesis A model of human evolution proposing that adaptations for feeding on small food items, such as seeds, grasses and tubers, provided the selection pressures for the evolution of hominid biological characteristics.

sexual dimorphism Differences between the sexes in physical appearance, behavioral performance, and psychological characteristics.

skeletal age A measure of biological maturation (as distinguished from chronological age) based on stages of formation of the bones.

socioeconomic status An indicator, often defined by measures of occupation and education, used as a proxy for the general quality of the environment for growth and development of an individual.

transformation grids A method developed by D'Arcy Thompson (based on drawings of the artist Albrecht Dürer) to describe two-dimensional changes in growth and form, both within and between species of organisms.

velocity curve A graphic representation of the rate of growth of an individual over time.

Literature cited

Note: In this list works by two joint authors are listed in alphabetical order by first author's name and then second author's name. Works by three or more joint authors, referenced in the text as e.g., 'Smith *et al.* 1970', 'Smith *et al.* 1980,' are listed alphabetically by the first author's name and, then, chronologically by date of publication.

Able, E. L. (1982). Consumption of alcohol during pregnancy: a review of effects on growth and development of offspring. *Human Biology*, **54**, 421–53.

Albright, F., Smith, P. H. & Fraser, R. (1942). A syndrome characterized by primary ovarian insufficiency and decreased stature: report on 11 cases with a digression on hormonal control of auxillary and pubic hair. *American Journal of Medical Science*, **204**, 625.

Alexander, R. D., Hoogland, J. L., Howard, R. D., Noonan, K. M. & Sherman, P. W. (1978). Sexual dimorphisms and breeding systems in pinnipeds, ungulates, primates, and humans. In *Evolutionary Biology and Human Social Behavior: An Anthropological Perspective*, ed. N. A. Chagnon & W. Irars, pp. 402–35. North Scituate, Massachusetts: Duxbury.

Alley, T. R. (1983). Growth-produced changes in body shape and size as determinants of perceived age and adult caregiving. *Child Development*, **54**, 241–8.

Altmann, J. (1980). *Baboon Mothers and Infants*. Cambridge: Harvard University Press.

Ammon, O. (1899). Zur Anthropologie der Badener. Jena, cited in Boas (1922).

Amoroso, E. C. (1961). Histology of the placenta. *British Medical Bulletin*, **17**, 81–8.

Anderson, D. C. (1980). The adrenal androgen-stimulating hormone does not exist. *Lancet*, **2**, 454–6.

Asayama, S. (1975). Adolescent sex development and adolescent behavior in Japan. *Journal of Sex Research*, **11**, 91–112.

Ashcroft, M. T. & Lovell, H. A. (1964). Heights and weights of Jamaican children of various racial origins. *Tropical and Geographical Medicine*, **4**, 346–53.

Ashcroft, M. T., Heneage, P. & Lovell, H. A. (1966). Heights and weights of Jamaican schoolchildren of various ethnic groups. *American Journal of Physical Anthropology*, **24**, 35–44.

Backman, G. (1934). Das Wachstum der Korperlange des Menchen. *Kunglicke Svenska Verenskapsakademiens Handlingar*, **14**, 145.

Bailey, S. M. & Garn, S. M. (1979). Socioeconomic interactions with physique and fertility. *Human Biology*, **51**, 317–33.

Bailey, S. M., Gershoff, S. N., McGandy, R. B., Nondasuta, A., Tantiwongse, P., Suttapreyasri, D., Miller, J. & McCree, P. (1984). A longitudinal study of growth and maturation in rural Thailand. *Human Biology*, **56**, 539–57.

228

Baker, P., ed. (1977). *The Biology of High Altitude Peoples*. Cambridge: Cambridge University Press.

Bakwin, H. (1942). Loneliness in infants. *American Journal of Disease of Children*, **63**, 30–40.

Barinaga, M., Bilezikjian, L. M., Vale, W. W., Rosenfeld, M. G. & Evans, R. M. (1985). Independent effects of growth hormone releasing factor on growth hormone release and gene transcription. *Nature*, **314**, 279–81.

Barnicot, N. A. (1977). Biological variation in modern populations. In *Human Biology*, 2nd edn, ed. G. A. Harrison, J. S. Weiner, J. M. Tanner & N. A. Barnicot, pp. 181–298. Oxford: Oxford University Press.

Bartholomew, G. A. & Birdsell, J. B. (1953). Ecology and the proto hominids. *American Anthropologist*, **55**, 481–98.

Baughan, B., Brault-Dubuc, M., Demirjian, A. & Gagnon, G. (1980). Sexual dimorphism in body composition changes during the pubertal period: as shown by French–Canadian children. *American Journal of Physical Anthropology*, **52**, 85–94.

Bayley, N. & Pinneau, S. R. (1952). Tables for predicting adult height from skeletal age: revised standards for use with the Greulich-Pyle hand standards. *Journal of Pediatrics*, **40**, 423–41.

Beach, F. (1974). Human sexuality and evolution. In *Human Evolution: Biosocial Perspectives*, ed. S. L. Washburn & E. R. McCown, pp. 123–53. Menlo Park: Benjamin/Cummings.

Beall, C. M. (1982). A comparison of chest morphology in high altitude Asian and Andean populations. *Human Biology*, **54**, 145–63.

Beall, C. M. & Reichsman, A. B. (1984). Hemoglobin levels in a Himalayan high altitude population. *American Journal of Physical Anthropology*, **63**, 301–6.

Beall, C. M., Baker, P. T., Baker, T. S. & Haas, J. D. (1977). The effects of high altitude on adolescent growth in southern Peruvian Amerindians. *Human Biology*, **49**, 109–24.

Beck, B. B. (1980). *Animal Tool Behavior*. New York: Garland.

Behar, M. (1977). Protein-caloric deficits in developing countries. *Annals of the New York Academy of Sciences*, **300**, 176.

Bercu, B. B., Lee, B. C., Spiliotis, B. E., Pineda, J. L., Derman, D. W., Hoffman, H. J., Brown, T. J. & Sachs, H. C. (1983). Male sexual development in the monkey. I. Cross-sectional analysis of pulsatile hypothalamic–pituitary–testicular function. *Journal of Clinical Endocrinology and Metabolism*, **56**, 1214–26.

Berkey, C. S., Reed, R. B. & Valadian, I. (1983). Midgrowth spurt in height of Boston children. *Annals of Human Biology*, **10**, 25–30.

Bertalanffy, L. von (1960). Principles and theory of growth. In *Fundamental Aspects of Normal and Malignant Growth*, ed. W. N. Nowinski, pp. 137–259. Amsterdam: Elsevier.

Bielicki, T. (1975). Interrelationships between various measures of maturation rate in girls during adolescence. *Studies in Physical Anthropology*, **1**, 51–64.

Bielicki, T. & Charzewski, J. (1983). Body height and upward social mobility. *Annals of Human Biology*, **10**, 403–8.

Bielicki, T. & Welon, Z. (1982). Growth data as indicators of social inequalities: The case of Poland. *Yearbook of Physical Anthropology*, **25**, 153–67.

Bielicki, T., Szczotke, H. & Charzewski, J. (1981). The influence of three socio-economic factors on body height in Polish military conscripts. *Human Biology*, **53**, 543–55.

Bielicki, T., Koniarek, J. & Malina, R. M. (1984). Interrelationships among certain measures of growth and maturation rate in boys during adolescence. *Annals of Human Biology*, **11**, 201–10.

Bierman, J. M., Siegel, E., French, F. E. & Simonian, K. (1965). Analysis of the outcome of all pregnancies in a community. *American Journal of Obstetrics and Gynecology*, **91**, 37–45.

Billewicz, W. Z. (1967). A note on body weight and seasonal variation. *Human Biology*, **39**, 241–50.

Billewicz, W. Z. & McGregor, I. A. (1982). A birth-to-maturity longitudinal study of heights and weights in two West African (Gambian) villages. *Annals of Human Biology*, **9**, 309–20.

Billewicz, W. Z., Fellows, H. M. & Thompson, A M. (1981). Pubertal changes in boys and girls in Newcastle upon Tyne. *Annals of Human Biology*, **8**, 211–19.

Bindon, J. R. & Baker, P. T. (1985). Modernization, migration and obesity among Samoan adults. *Annals of Human Biology*, **12**, 67–76.

Birdsell, J. B. (1979). Ecological influences on Australian Aboriginal social organization. In *Primate Ecology and Human Origins*, ed I. S. Bernstein & E. O. Smith, pp. 117–51. New York: Garland.

Blank, M. S., Panerai, S. & Griesan, H. (1979). Opioid peptides modulate luteinizing hormone secretion during sexual maturation. *Science*, **203**, 1129–31.

Boas, F. (1892). The growth of children, II. *Science*, **19**, 281–2.

Boas, F. (1912). Changes in the bodily form of descendants of immigrants. *American Anthropologist*, **14**, 530–63.

Boas, F. (1922). Report on an anthropometric investigation of the population of the United States. *Journal of the American Statistical Association*, **18**, 181–209.

Boas, F. (1930). Observations on the growth of children. *Science*, **72**, 44–8.

Boas, F. (1940). *Race, Language, & Culture*. New York: Free Press.

Bock, R. D. (1986). Unusual growth patterns in the Fels data. In *Human Growth: A Multidisciplinary Review*, ed. A. Demirjian, pp. 69–84, London: Taylor & Francis.

Bock, R. D. & Thissen, D. M. (1976). Fitting multi-component models for growth in stature. *Proceedings of the Ninth International Biometric Conference*, **1**, 431–42.

Bock, R. D. & Thissen, D. (1980). Statistical problems of fitting individual growth curves. In *Human Physical Growth and Maturation, Methodologies and Factors*, ed. F. E. Johnston, A. F. Roche & C. Susanne, pp. 265–90. New York: Plenum.

Bock, R. D., Wainer, H., Peterson, A., Thissen, J. M. & Roche, A. (1973). A parameterization for individual human growth curves. *Human Biology*, **45**, 63–80.

Bogin, B. (1979). Monthly changes in the gain and loss of growth in weight of children living in Guatemala. *American Journal of Physical Anthropology*, **51**, 287–92.

Bogin, B. (1980). Catastrophe theory model for the regulation of human growth. *Human Biology*, **52**, 215–27.

Bogin, B. (1986). Auxology and anthropology. *Reviews in Anthropology*, **13**, 7–13.

Bogin, B. (1988). Rural-to-urban migration. In *Biological Aspects of Human Migration*, ed. C. G. N. Mascie-Taylor & G. W. Lasker, pp. 90–129. Cambridge: Cambridge University Press.

Bogin, B. A. (1977). *Periodic Rhythm in the Rates of Growth in Height and Weight of Children and its Relation to Season of the Year*. Ph.D. Dissertation. Ann Arbor: University Microfilms.

Bogin, B. A. (1978). Seasonal pattern in the rate of growth in height of children living in Guatemala, *American Journal of Physical Anthropology*, **49**, 205–10.

Bogin, B. & MacVean, R. B. (1978). Growth in height and weight of urban Guatemalan primary school children of high and low socioeconomic class. *Human Biology*, **50**, 477–88.

Bogin, B. & MacVean, R. B. (1981a). Body composition and nutritional status of urban Guatemalan children of high and low socioeconomic class. *American Journal of Physical Anthropology*, **55**, 543–51.

Bogin, B. & MacVean, R. B. (1981b). Bio-social effects of migration on the development of families and children in Guatemala. *American Journal of Public Health*, **71**, 1373–7.

Bogin, B. & MacVean, R. B. (1981c). Nutritional and biological determinants of body fat patterning in urban Guatemalan children. *Human Biology*, **53**, 259–68.

Bogin, B. & MacVean, R. B. (1982). Ethnic and secular influences on the size and maturity of seven year old children living in Guatemala City. *American Journal of Physical Anthropology*, **59**, 393–8.

Bogin, B. & MacVean, R. B. (1983). The relationship of socioeconomic status and sex to body size, skeletal maturation, and cognitive status of Guatemala City schoolchildren. *Child Development*, **54**, 115–28.

Bogin, B. & MacVean, R. B. (1984). Growth status of non-agrarian, semi-urban living Indians in Guatemala. *Human Biology*, **56**, 527–38.

Bogin, B. & Sullivan, T. (1986). Socioeconomic status, sex, age, and ethnicity as determinants of body fat distribution for Guatemalan children. *American Journal of Physical Anthropology*, **69**, 527–35.

Bolk, L. (1926). *Das Problem der Menschwerdung*. Jena: Gustav Fischer.

Bookstein, F. C. (1978). *The Measurement of Biological Shape and Shape Change*. New York: Springer-Verlag.

Borkan, G. A., Hults, D. E., Cardarelli, J. & Burrows, B. A. (1982). Comparison of ultrasound and skinfold measurements in assessment of subcutaneous and total fatness. *American Journal of Physical Anthropology*, **58**, 307–13.

Borms, J. (1984). Preface. In *Human Growth and Development*, ed. J. Borms *et al.*, pp. v–vii. New York: Plenum.

Borms, J., Hauspie, R., Sand, C., Susanne, C., Hebbelinck, M., eds. (1984). *Human Growth and Development*. New York: Plenum.

Bowlby, J. (1969). *Attachment and Loss*. New York: Basic Books.

Boyar, R., Finkelstein, J., Roffwarg, H., Kapen, S., Weitzman E. & Hellman, L. (1972). Synchronization of augmented luteinizing hormone secretion with sleep during puberty. *New England Journal of Medicine*, **287**(12), 582–6.

Boyar, R., Finkelstein, J., Roffwarg, H., Kapen, S., Weitzman, E. & Hellman, L. (1973). Twenty-four hour luteinizing hormone and follicle-stimulating hormone secretory patterns in gonadal dysgenesis. *Journal of Clinical Endocrinology and Metabolism*, **37**, 521–4.

Boyar, R. M., Rosenfeld, R. S., Kapan, S., Finkelstein, J. W., Roffwarg, H. P., Weitzman, E. D. & Hellman, L. (1974). Simultaneous augmented secretion of luteinizing hormone and testosterone during sleep. *Journal of Clinical Investigation*, **54**, 609–18.

Boyar, R. M., Wu, R. H. K., Roffwarg, H., Kapen, S., Weitzman, E. D., Hellman, L. & Finkelstein, J. W. (1976). Human puberty–24 hour estradiol patterns in prepubertal girls. *Journal of Clinical Endocrinology and Metabolism*, **43**, 1418–21.

Boyce, A. J., ed. (1984). *Migration and Mobility*. London: Taylor & Francis.

Boyd, E. (1981). *Origins of the Study of Human Growth*, ed. B. S. Savara & J. F. Schilke. Eugene: University of Oregon Press.

Brain, C. K. (1981). *The Hunters or the Hunted? An Introduction to African Cave Taphonomy*. Chicago: University of Chicago Press.

Britten, R. J. & Davidson, E. H. (1969). Gene regulation for higher cells: a theory. *Science*, **165**, 349–57.

Brody, S. (1945). *Bioenergetics and Growth*. New York: Reinhold Publishing Co.

Bromage, T. G. & Dean, M. C. (1985) Re-evaluation of the age at death of immature fossil hominids. *Nature*, **317**, 525–7.

Brook, C. G. D., Gasser, T., Werder, E. A., Prader, A. & Vanderschuren-Ludewyks, M. A. (1977). Height correlations between parents and mature offspring in normal subjects and in subjects with Turner's and Klinefelter's and other syndromes. *Annals of Human Biology*, **4**, 17–22.

Brown, F., Harris, J., Leakey, R. & Walker, A. (1985). Early *Homo erectus* skeleton from west Lake Turkana, Kenya. *Nature*, **316**, 788–92.

Brown, T. (1983). The Preece–Baines growth function demonstrated by personal computer: a teaching and research aid. *Annals of Human Biology*, **10**, 487–9.

Brown, T. & Townsend, G. C. (1982). Adolescent growth in height of Australian Aboriginals analysed by the Preece–Baines function: a longitudinal study. *Annals of Human Biology*, **9**, 495–505.

Brozek, J. (1960). The measurement of body composition. In *A Handbook of Anthropometry*, ed. M. F. Ashley Montagu, pp. 78–120. Springfield, Illinois: Charles C. Thomas.

Brundtland, G. H. & Walløe, L. (1973). Menarcheal age in Norway: Halt in trend towards earlier maturation. *Nature*, **241**, 478–9.

Brundtland, G. H., Liestøl, K. & Walløe, L. (1980). Height, weight, and menarcheal age of Oslo schoolchildren during the last 60 years. *Annals of Human Biology*, **7**, 307–22.

Brush, G., Boyce, A. J. & Harrison, G. A. (1983). Associations between anthropometric variables and reproductive performance in a Papua New Guinea highland population. *Annals of Human Biology*, **10**, 223–34.

Buffon, G. (1977). *Histoire Naturelle*. Fourth Supplement, Paris.

Burt, C. (1937). *The Backward Child*. New York: Appleton–Century.

Byard, P. J., Siervogel, R. M. & Roche, A. F. (1983). Familial correlations for serial measurements of recumbent length and stature. *Annals of Human Biology*, **10**, 281–93.

Cameron, N., Tanner, J. M. & Whitehouse, R. H. (1982). A longitudinal analysis of the growth of limb segments in adolescence. *Annals of Human Biology*, **9**, 211–20.

Campbell, B. (1985). *Human Evolution*, 3rd edn. New York: Aldine.

Case, T. J. (1978). On the evolution and adaptive significance of postnatal growth rates in terrestrial vertebrates. *Quarterly Review of Biology*, **53**, 243–82.

Chang, K. S. F., Ng, P. H., Lee, M. M. C. & Chan, S. J. (1966). Sexual maturation of Chinese boys in Hong Kong. *Pediatrics*, **37**, 804–11.

Cheek, D. B. (1968). Muscle cell growth in normal children. In *Human Growth*, ed. D. B. Cheek, pp. 337–51. Philadelphia: Lea & Febiger.

Chow, B. (1974). Effect of maternal dietary protein on anthropometric and behavioral development of the offspring. In *Nutrition and Malnutrition: Identification and Measurement*, ed. A. F. Roche & F. Falkner, pp. 189–219, New York: Plenum.

Clark, P. J. & Spuhler, J. N. (1959). Differential fertility in relation to body dimensions. *Human Biology*, **31**, 121–37.

Clegg, E. J. (1982). The influence of social, geographical and demographic factors on the size of 11–13 year old children from the Isle of Lewis, Scotland. *Human Biology*, **54**, 93–109.

Clegg, E. J., Pawson, I. G., Ashton, E. H. & Flinn, R. M. (1972). The growth of children at different altitudes in Ethiopia. *Philosophical Transactions of the Royal Society of London*, **264B**, 403–37.

Clemmons, D. R. & Van Wyk, J. J. (1984). Factors controlling blood concentration of somatomedin-C. *Journal of Clinical Endocrinology and Metabolism*, **13**, 113–43.

Clemmons, D. R., Klibanski, A., Underwood, L. E., McArthur, J. W., Ridgway, E. C., Beitins, I. Z. & Van Wyk, J. J. (1981). Reduction of plasma immunoreactive somatomedin-C during fasting in humans. *Journal of Clinical Endocrinology and Metabolism*, **53**, 1247–50.

Coelho, A. M. Jr. (1985). Baboon dimorphism: growth in weight, length and adiposity from birth to 8 years of age. In *Nonhuman Primate Models for Human Growth*, ed. E. S. Watts, pp. 125–59. New York: Alan R. Liss.

Count, E. W. (1943). Growth patterns of the human physique: an approach to kinetic anthropometry. *Human Biology*, **15**, 1–32.

Cousins, R. J. & Deluca, H. F. (1972). Vitamin D and bone. In *The Biochemistry and Physiology of Bone*, ed. G. H. Bourne, pp. 281–335. New York: Academic Press.

Crognier, E. (1981). Climate and anthropometric variations in Europe and the Mediterranean area. *Annals of Human Biology*, **8**, 99–107.

Cutler, G. B. Jr., Glenn, M., Bush, M., Hodgen, G. D., Graham, C. E. & Loriaux, D. L. (1978). Adrenarche: a survey of rodents, domestic animals, and primates. *Endocrinology*, **103**, 2112–18.

Damon, A. & Thomas, R. B. (1967). Fertility and physique – height, weight and ponderal index. *Human Biology*, **39**, 5–13.

Davies, P. S. W., Jones, P. R. M. & Norgan, N. G. (1986). The distribution of subcutaneous and internal fat in man. *Annals of Human Biology*, **13**, 189–92.

Dawkins, R. (1976). *The Selfish Gene*. Oxford: Oxford University Press.

Dean, M. C. & Wood, B. A. (1981). Developing pongid dentition and its use for ageing individual crania in comparative cross-sectional growth studies. *Folia Primatologia*, **36**, 111–27.

D'Ercole, A. J. & Underwood, L. E. (1986). Regulation of fetal growth by hormones and growth factors. In *Human Growth*, Vol. 1, 2nd edn, ed. F. Falkner & J. M. Tanner, pp. 327–38. New York: Plenum.

Deming, J. (1957). Application of the Gompertz curve to the observed pattern of growth in length of 48 individual boys and girls during the adolescent cycle of growth. *Human Biology*, **29**, 83–122.

Demirjian, A. (1986). Dentition. In *Human Growth*, Vol. 2, eds. F. Falkner & J. M. Tanner, pp. 269, 298, New York: Plenum.

Devi, R. M., Kumari, J. R. & Srikumari, C. R. (1985). Fertility and mortality differences in relation to maternal body size. *Annals of Human Biology*, **12**, 479–84.

DeVilliers, H. (1971). A study of morphological variables in urban and rural Venda male populations. In *Human Biology of Environmental Change*, ed. D. J. M. Vorster. London: International Biological Program.

Dittus, W. P. J. (1977). The social regulation of population density and age–sex distribution in the Toque Monkey. *Behaviour*, **63**, 281–322.

Dobzhansky, T. (1962). *Mankind Evolving*. New Haven: Yale University Press.

Dobzhansky, T. (1973). Nothing in biology makes sense except in the light of evolution. *American Biology Teacher*, **35**, 125–9.

Donaldson, H. H. (1908). *The Growth of the Brain*. New York: Charles Scribner's Sons.

Draper, P. (1976). Social and economic constraints on child life among the !Kung. In *Kalahari Hunters-Gatherers*, ed. R. B. Lee & I. DeVore, pp. 199–217. Cambridge: Cambridge University Press.

Dubos, R. (1965). *Man Adapting*, New Haven: Yale University Press.

Eccles, J. C. (1979). *The Understanding of the Brain*. New York: McGraw-Hill.

Eigenmann, J. E., Patterson, D. F. & Froesch, E. R. (1984) Body size parallels insulin-like growth factor I levels but not growth hormone secretory capacity. *Acta Endocrinologica*, **106**, 448–53.

Elkin, A. P. (1964). *The Australian Aborigines*. New York: Doubleday/Anchor.

Enlow, D. H. (1963). *Principles of Bone Remodeling*. Springfield, Illinois: Charles C. Thomas.

Enlow, D. H. (1976). The remodeling of bone. *Yearbook of Physical Anthropology*, **20**, 19–34.

Eveleth, P. B. & Tanner, J. M. (1976). *World-Wide Variation in Human Growth*. Cambridge: Cambridge University Press.

Falkner, F. (1966). General considerations in human development. In *Human Development*, ed. F. Faulkner, pp. 10–39. Philadelphia: Saunders.

Falkner, F. (1978). Implications for growth in human twins. In *Human Growth*, Vol. 1, ed. F. Falkner & J. M. Tanner, pp. 397–413. New York: Plenum.

Fischbein, S. (1977). Onset of puberty in MZ and DZ twins. *Acta Geneticae Medicae et Gemellologiae*, **26**, 151–8.

Fogel, R. W. (1986). Physical growth as a measure of the economic wellbeing of populations: The eighteenth and nineteenth centuries. In *Human Growth*, Vol. 3, 2nd edn, ed. F. Falkner & J. M. Tanner, pp. 263–81. New York: Plenum.

Frisancho, A. R. (1977). Human growth and development among high-altitude populations. In *The Biology of High Altitude Peoples*, ed. P. Baker, pp. 117–71. Cambridge: Cambridge University Press.

Frisancho, A. R. (1979). *Human Adaptation: A Functional Interpretation*. St Louis: Mosby.

Frisancho, A. R. (1981). New norms of upper limb fat and muscle areas for assessment of nutritional status. *American Journal of Clinical Nutrition*, **34**, 2540–5.

Frisancho, A. R. & Baker, P. T. (1970). Altitude and growth: A study of the patterns of physical growth of a high altitude Peruvian Quechua population. *American Journal of Physical Anthropology*, **32**, 279–92.

Frisancho, A. R., Garn, S. M. & Ascoli, W. (1970). Childhood retardation resulting in reduction of adult body size due to to lesser adolescent skeletal delay. *American Journal of Physical Anthropology*, **33**, 325–36.

Frisancho, R. A., Sanchez, J., Pallardel, D. & Yanez, L. (1973). Adaptive significance of small body size under poor socioeconomic conditions in Southern Peru. *American Journal of Physical Anthropology*, **39**, 255–62.

Frisancho, A. R., Borkan, G. A. & Klayman, J. F. (1975). Pattern of growth of Lowland and Highland Peruvian Quechua of similar genetic composition. *Human Biology*, **47**, 233–43.

Frisancho, R. A., Guire, K., Babler, W., Borkan, G. & Way, A. (1980). Nutritional influence of childhood development and genetic control of adolescent growth of Quechuas and Mestizos from the Peruvian Lowlands. *American Journal of Physical Anthropology*, **52**, 367–75.

Frisancho, R. A., Matos, J., Leonard, W. R. & Yaroch, L. A. (1985). Developmental and nutritional determinants of pregnancy outcome among teenagers. *American Journal of Physical Anthropology*, **66**, 247–61.

Frisch, R. E. (1974). Critical weight at menarche, initiation of the adolescent growth spurt and control of puberty. In *Control of the Onset of Puberty*, ed. M. M. Grumbach, G. D. Grace & F. F. Mayer, pp. 403–23. New York: Wiley.

Frisch, R. E. & Revelle, R. (1970). Height and weight at menarche and a hypothesis of critical body weights and adolescent events. *Science*, **169**, 397–8.

Frisch, R. E., Wyshak, G. & Vincent, L. (1980). Delayed menarche and amenorrhea in ballet dancers. *New England Medical Journal*, **303**, 17–19.

Froment, A. & Hiernaux, J. (1984). Climate-associated anthropometric variation between populations of the Niger bend. *Annals of Human Biology*, **11**, 189–200.

Fulwood, R., Abraham, S. & Johnson, C. (1981). *Height and Weight of Adults Ages 18–74 Years by Socioeconomic and Geographic Variables*. Vital and Health Statistics, Series 11, No. 224, DHEW Pub. No. (PHS) 81–1674. Washington, DC: US Government Printing Office.

Garn, S. M. (1985). Smoking and human biology. *Human Biology*, **57**, 505–23.

Garn, S. M. & Bailey, S. M. (1978). Genetics of the maturational processes. In *Human Growth*, Vol. 1, ed. F. Falkner & J. M. Tanner, pp. 307–30. New York: Plenum.

Garn, S. M. & Clark, D. C. (1975). Nutrition, growth, development, and maturation: Findings from the Ten-State Nutrition Survey of 1968–1970. *Pediatrics*, **56**, 300–19.

Garn, S. M. & Clark, D. C. (1976). Problems in the nutritional assessment of black individuals. *American Journal of Public Health*, **66**, 262–7.

Garn, S. M. & Petzold, A. S. (1983). Characteristics of the mother and child in teenage pregnancy. *American Journal of Diseases of Children*, **137**, 365–8.

Garn, S. M. & Rohmann, C. G. (1962). X-linked inheritance of developmental timing in man. *Nature*, **196**, 695–6.

Garn, S. M. & Rohman, C. G. (1966) Interaction of nutrition and genetics in the timing of growth and development. *Pediatric Clinics of North America*, **13**, 353–79.

Garn, S. M., Owen, G. M. & Clark, D. C. (1974). Ascorbic acid: The vitamin of affluence. *Ecology of Food and Nutrition*, **3**, 151–3.

Garn, S. M., Bailey, S. M. & Higgins, I. T. T. (1976). Fatness similarities in adopted pairs. *American Journal of Clinical Nutrition*, **29**, 1067–8.

Garn, S. M., Shaw, H. A. & McCabe, K. D. (1977). Effect of socioeconomic status and race on weight defined and gestational prematurity in the United States. In *The Epidemiology of Prematurity*, ed. D. W. Reed & F. J. Stanley, pp. 127–43. Baltimore: Urban & Scharzenberg.

Garn, S. M., Cole, P. E. & Bailey, S. M. (1979). Living together as a factor in family-line resemblances. *Human Biology*, **51**, 565–87.

Garn, S. M., Pesick, S. D. & Pilkington, J. J. (1984). The interaction between prenatal and socioeconomic effects on growth and development in childhood. In *Human Growth and Development*, ed. J. Borms, R. Hauspie, A. Sand, C. Susanne & M. Hebbelinck, pp. 59–70. New York: Plenum.

Garrow, J. S. & Pike, M. C. (1967). The long term prognosis of severe infantile malnutrition. *Lancet*, **1**, 1–4.

Gasser, T., Kohler, W., Muller, H. G., Kneip, A., Largo, R., Molinari, L. & Prader, A. (1984). Velocity and acceleration of height growth using kernel estimation. *Annals of Human Biology*, **11**, 397–411.

Gasser, T., Kohler, W., Muller, H. G., Largo, R., Molinari, L. & Prader, A. (1985). Human height growth: Correlational and multivariate structure of velocity and acceleration. *Annals of Human Biology*, **12**, 501–15.

Gavan, J. A. (1953). Growth and development of the chimpanzee, a longitudinal and comparative study, *Human Biology*, **25**, 93–143.

Gavan, J. A. (1971). Longitudinal postnatal growth in the chimpanzee. In *The Chimpanzee*, Vol. 4, ed. G. Bourne, pp. 46–102. Basel: Karger.

Gavan, J. A. (1982). Adolescent growth in non-human primates: an introduction. *Human Biology*, **54**, 1–5.

Goldstein, H. (1984). Current developments in the design and analysis of growth studies. In *Human Growth and Development*, ed. J. Borms, R. Hauspie, A. Sand, C. Susanne & M. Hebbelinck, pp. 733–52. New York: Plenum.

Goldstein, M. S. (1943). *Demographic and Bodily Changes in Descendants of Mexican Immigrants*. Austin: Institute of Latin American Studies.

Gompertz, B. (1825). On the nature of the function expressive of the law of human mentality. *Philosophical Transaction of the Royal Society*, **115**, 513–83.

Goodall, J. (1983). Population dynamics during a 15-year period in one community of free-living chimpanzees in the Gombe National Park, Tanzania. *Zietschrift für Tierpsychologie*, **61**, 1–60.

Goss, R. (1964). *Adaptive Growth*. New York: Academic Press.

Goss, R. (1978). *The Physiology of Growth*. New York: Academic Press.

Goss, R. (1986). Modes of growth and regeneration. In *Human Growth*, Vol. 1, 2nd edn, ed. F. Falkner & J. M. Tanner, pp. 3–26. New York: Plenum.

Gould, J. B. (1986). The low birth weight infant. In *Human Growth*, Vol. 1, 2nd edn, ed. F. Falkner & J. M. Tanner, pp. 391–413. New York: Plenum.

Gould, S. J. (1977). *Ontogeny and Phylogeny*. Cambridge: Belknap.

Gould, S. J. (1979). Mickey Mouse meets Konrad Lorenz. *Natural History*, **88**(4), 30–6.

Gould, S. J. (1981). *The Mismeasure of Man*. New York: Norton.

Grant, D. B., Hambley, J., Becker, D., Pimstone, B. L. (1973). Reduced sulphation factor in undernourished children. *Archives of Disease in Childhood*, **48**, 596–600.

Green, H., Morikawa, M. & Nixon, T. (1985). A duel effector theory of growth-hormone action. *Differentiation*, **29**, 195–8.

Green, W. H., Campbell, M. & David, R. (1984). Psychosocial dwarfism: a critical review of the evidence. *Journal of the American Academy of Child Psychiatry*, **23**, 39–48.

Greene, L. S. (1973). Physical growth and development, neurological maturation and behavioral functioning in two Ecuadorian Andean communities in which goiter is endemic. *American Journal of Physical Anthropology*, **38**, 119–34.

Gregor, T. (1979). Short people. *Natural History*, February, 14–23.

Greulich, W. W. (1976). Some secular changes in the growth of American-born and native Japanese children. *American Journal of Physical Anthropology*, **45**, 553–68.

Greulich, W. W. & Pyle, S. I. (1959). *Radiographic Atlas of Skeletal Development of the Hand and Wrist*, 2nd edn. Stanford: Stanford University Press.

Grumbach, M. M. (1978). The central nervous system and the onset of puberty. In Human Growth, Vol. 2, ed. F. Falkner & J. M. Tanner, pp. 215–38. New York: Plenum.

Grumbach, M. M., Roth, J. C., Kaplan, S. L. & Kelch, R. P. (1974). Hypothalmic–pituitary regulation of puberty in man: evidence and concepts derived from clinical research. In *Control of the Onset of Puberty*, ed. M. M. Grumbach, G. D. Grave & F. E. Mayer, pp. 115–66. New York: Wiley.

Gupta, R. & Basu, A. (1981). Variations in body dimensions in relation to altitude among the Sherpas of the eastern Himalayas. *Annals of Human Biology*, **8**, 145–51.

Gurney, J. M. & Jelliffe, D. B. (1973). Arm anthropometry in nutritional assessment: nomogram for the rapid calculation of muscle circumference and cross-sectional muscle and fat areas. *American Journal of Clinical Nutrition*, **26**, 912–15.

Guthrie, H. A. (1986). *Introductory Nutrition*. St Louis: Times Mirror/Mosby.

Habicht, J.-P., Yarbrough, C., Martorell, R., Malina, R. M. & Klein, R. E. (1974). Height and weight standards for preschool children. *Lancet*, **1**, 611–15.

Haddad, J. G. & Hahn, T. J. (1973). Natural and synthetic sources of circulating 25-hydroxy vitamin D in man. *Nature*, **224**, 515–16

Hallowell, A. I. (1960). Self, society and culture in phylogenetic perspective. In *The Evolution of Man*, ed. S. Tax, pp. 309–71. Chicago: University of Chicago Press.

Hamill, P. V. V., Johnston, F. E. & Lemshow, S. (1973). *Body weight, stature, and sitting height: white and Negro youths 12–17 years, United States*. DHEW Publication No. (HRA) 74–1698. Washington, DC: US Government Printing Office.

Hamill, P. V. V., Johnson, C. L., Reed, R. B. & Roche, A. F. (1977). *NCHS Growth Curves for Children Births–18 years, United States*. DHEW Publications, (PHS) 78–1650. Washington, DC: US Government Printing Office.

Hamilton, W. J. & Mossman, H. (1972) *Human Embryology: Prenatal Development of Form and Function*, 4th edn. Cambridge: Heffer & Sons.

Hansman, C. (1970). Anthropometry and related data: anthropometry, skinfold thickness measurements. In *Human Growth and Development*, ed. R. W. McCammon, pp. 101–54. Springfield, Illinois: Charles C. Thomas.

Harsha, D. W., Voors, A. W. & Berenson, G. S. (1980). Racial differences in subcutaneous fat patterns in children aged 7–15 years. *American Journal of Physical Anthropology*, **53**, 333–7.

Harvey, P. H. & Zammuto, R. M. (1985). Patterns of mortality and age at first reproduction in natural populations of mammals. *Nature*, **315**, 319–20.

Hass, J. D., Frongillo, E. A. Jr., Stepick, C. D., Beard, J. L. & Hurtado, G. (1980). Altitude, ethnic and sex difference in birth weight and length in Bolivia. *Human Biology*, **52**, 459–77.

Hauspie, R. C., Das, S. R., Preece, M. A. & Tanner, J. M. (1980a). A longitudinal study of the growth in height of boys and girls of West Bengal (India) aged six months to 20 years. *Annals of Human Biology*, **7**, 429–41.

Hauspie, R. C., Wachholder, A., Baron, G., Cantrine, F., Susanne, C. & Graffar, M. (1980b). A comparative study of the fit of four different functions to longitudinal data of growth in height of Belgian girls. *Annals of Human Growth*, **7**, 347–58.

Hayflick, L. (1980). The cell biology of human aging. *Scientific American*, **242**, 58–65.

Hediger, M. L. & Katz, S. H. (1986). Fat patterning, overweight, and adrenal androgen interactions in black adolescent females. *Human Biology*, **58**, 585–600.

Hiernaux, J. (1974). *The People of Africa*. London: Weidenfeld & Nicolson.

Higham, E. (1980). Variations in adolescent psychohormonal development. In *Handbook of Adolescent Psychology*, ed. J. Adelson, pp. 472–94. New York: Wiley.

Hintz, R. L., Suskind, R., Amatayakerl, K., Thanangkul, O., & Olson, R. (1978). Plasma somatomedin and growth hormone values in children with protein–calorie malnutrition. *Journal of Pediatrics*, **92**, 153–6.

Hirsch, M., Shemesh, J., Modan, M. & Lunenfeld, B. (1979). Emission of spermatoza. Age of onset. *International Journal of Andrology*, **2**, 289–98.

Holliday, M. A. (1986). Body composition and energy needs during growth. In *Human Growth*, Vol. 2, 2nd edn, F. Falkner & J. M. Tanner, pp. 101–17. New York: Plenum.

Hoshi, H. & Kouchi, M. (1981). Secular trend of the age at menarche of Japanese girls with special regard to the secular acceleration of the age at peak height velocity. *Human Biology*, **53**, 593–8.

Howe, P. E. & Schiller, M. (1952). Growth responses of the school child to changes in diet and environmental factors. *Journal of Applied Physiology*, **5**, 51–61.

Howell, N. (1976). The population of the Dobe area !Kung. In *Kalahari Hunter-Gatherers*, ed. R. B. Lee & I. DeVore, pp. 137–57. Cambridge: Cambridge University Press.

Howell, N. (1979). *Demography of the Dobe !Kung*. New York: Academic Press.

Hrdy, S. B. (1981). *The Woman That Never Evolved*. Cambridge: Harvard University Press.

Hulanicka, B. & Kotlarz, K. (1983). The final phase of growth in height. *Annals of Human Biology*, **10**, 429–34.

Hulse, F. (1969). Migration and cultural selection in human genetics. In *The Anthropologist*, ed. P. C. Biswas, pp. 1–21. Delhi, India: University of Delhi.

Hunt, E. E. (1966). The developmental genetics of man. In *Human Development*, ed. F. Falkner, pp. 76–122. Philadelphia: W. B. Saunders.

Hunt, E. E. Jr. & Heald, F. P. (1963). Physique, body composition, and sexual maturation in adolescent boys. *Annals of the New York Academy of Sciences*, **110**, 532–44.

Huxley, J. S. (1932). *Problems of Relative Growth*. London: Methuen; 2nd edn, 1972. New York: Dover.

Huxley, T. H. (1863). *Evidence as to Man's Place in Nature*. London: Williams & Norgate.

Illsley, R., Finlayson, A. & Thompson, B. (1963). The motivation and characteristics of internal migrants: a socio-medical study of young migrants in Scotland. *Milbank Memorial Fund Quarterly*, **41**, 115–44, 217–48.

Isaksson, O. G. P., Jansson, J.-O. & Gause, I. A. M. (1982). Growth hormone stimulates longitudinal bone growth directly. *Science*, **216**, 1237–8.

Isley, W. L., Underwood, L. E. & Clemmons, D. R. (1983). Dietary components that regulate serum somatomedin-C concentrations in humans. *Journal of Clinical Investigation*, **71**, 175–82.

Issac, G. (1978). The food sharing behavior of protohuman hominids. *Scientific American*, **238**(4), 90–108.

Itani, J. (1958). On the acquisition and propagation of a new food habit in the troop of Japanese monkeys at Takasakiyama. *Primates*, **1**, 131–48.

Jakacki, R. I., Kelch, R. P., Sander, S. E., Lloyd, J. S., Hopwood, N. J. & Marshall, J. C. (1982). Pulsatile secretion of luteinizing hormone in children. *Journal of Endocrinology and Metabolism*, **55**, 453–8.

Jelliffe, D. B. (1966). *The Assessment of the Nutritional Status of the Community*. WHO Monograph No. 53. Geneva: World Health Organization.

Jenkins, C. L. (1981). Patterns of growth and malnutrition among preschoolers in Belize. *American Journal of Physical Anthropology*, **56**, 169–78.

Jenkins, C. L., Orr–Ewing, A. K. & Heywood, P. F. (1984). Cultural aspects of early childhood growth and nutrition among the Amele of lowland Papua New Guinea. *Ecology of Food and Nutrition*, **14**, 261–75.

Jerison, H. S. (1973). *Evolution of the Brain and Intelligence*. New York: Academic Press.

Jerison, H. S. (1976). Paleoneurology and the evolution of mind. *Scientific American*, **234**(1), 90–101.

Johnson, C. L., Fulwood, R., Abraham, S. & Bryner, J. D. (1981). *Basic data on anthropometric measurements and angular measurements of the hip and knee joints for selected age groups 1–74 years of age*. DHHS Publication No. (PHS) 81–1669. Washington DC: US Government Printing Office.

Johnson, T. D. (1982). Selective costs and benefits in the evolution of learning. *Advances in the Study of Behaviour*, **12**, 65–106.

Johnston, F. E. (1974). Control of age at menarche. *Human Biology*, **46**, 159–71.

Johnston, F. E. (1980). Nutrition and growth. In *Human Physical Growth and Maturation*, ed. F. E. Johnston, A. Roche & C. Susanne, pp. 291–300. New York: Plenum.

Johnston, F. E. (1986). Somatic growth of the infant and preschool child. In *Human Growth*, Vol. 2, 2nd edn, ed. F. Falkner & J. M. Tanner, pp. 3–24. New York: Plenum.

Johnston, F. E. & Beller, A. (1976). Anthropometric evaluation of the body composition of black, white, and Puerto Rican newborns. *American Journal of Clinical Nutrition*, 29, 61.

Johnston, F. E., Borden, M. & MacVean, R. B. (1973). Height, weight and their growth velocities in Guatemalan private schoolchildren of high socio-economic class. *Human Biology* 45, 627–41.

Johnston, F. E., Hamill, P. V. V. & Lemeshow, S. (1974a). *Skinfold thickness of youths 12–17 years, United States*. NCHS, Vital and Health Statistics, Series II, No. 132. Washington, DC: US Government Printing Office.

Johnston, F. E., Hamill, P. V. V. & Lemeshow, S. (1974b). Skinfold thicknesses in a national probability sample of U.S. males and females 6 through 17 years. *American Journal of Physical Anthropology*, 40, 321–4.

Johnston, F. E., Dechow, P. C. & MacVean, R. B. (1975). Age changes in skinfold thickness among upper class school children of differing ethnic backgrounds residing in Guatemala. *Human Biology*, 47, 251–62.

Johnston, F. E., Wainer, H., Thissen, D. & MacVean, R. B. (1976). Hereditary and environmental determinants of growth in height in a longitudinal sample of children and youth of Guatemalan and European ancestry. *American Journal of Physical Anthropology*, 44, 469–76.

Johnston, F. E., Scholl, T. O., Newman, B. C., Cravioto, J. & De Licardie, E. R. (1980). An analysis of environmental variables and factors associated with growth failure in a Mexican village. *Human Biology*, 52, 627–37.

Johnston, F. E., Bogin, B., MacVean, R. B. & Newman, B. C. (1984). A comparison of international standards versus local reference data for the triceps and subscapular skinfolds of Guatemalan children and youth. *Human Biology*, 56, 157–71.

Johnston, F. E., Low, S. M., de Baessa, Y. & MacVean, R. B. (1985). Growth status of disadvantaged urban Guatemalan children of a resettled community. *American Journal of Physical Anthropology*, 68, 215–24.

Jolly, A. (1985). *The Evolution of Primate Behavior*, 2nd ed. New York: Macmillan.

Jolly, C. J. (1970). The seed eaters: a new model of hominid differentiation based on a baboon analogy. *Man*, 5, 5–26.

Jones, P. R. M. & Dean, R. F. A. (1956). The effects of Kwashiorkor on the development of the bones of the hand. *Journal of Tropical Pediatrics*, 2, 51.

Judd, H. L., Parker, D. C. & Yen, S. S. C. (1977). Sleep–wake pattern of LH and testosterone release in prepubertal boys. *Journal of Clinical Endocrinology and Metabolism*, 44, 865–9.

Kaplan, B. (1954). Environment and human plasticity. *American Anthropologist*, 56, 780–99.

Karlberg, J. (1985). The human growth curve decomposed into three additive and partly superimposed components: the FBP-model. In *Abstracts. IVth International Congress of Auxology*, p. 46. London: Taylor & Francis.

Katch, V. L., Campaigne, B., Freedson, P., Sady, S., Katch, F. I. & Behnke, A. R. (1980). Contribution of breast volume and weight to body fat distribution in females. *American Journal of Physical Anthropology*, 53, 93–100.

Katchadourian, H. (1977). *The Biology of Adolescence.* San Francisco: Freeman.

Katz, S. H., Hediger, M. L., Zemel, B. S. & Parks, J. S. (1985). Adrenal androgens, body fat and advanced skeletal age in puberty: new evidence for the relations of adrenarche and gonadarche in males. *Human Biology*, **57**, 401–13.

Kawai, M. (1965). Newly acquired precultural behavior of the natural troop of Japanese monkeys on Koshima Island. *Primates*, **6**, 1–30.

Kawamura, S. (1959). The process of sub-culture propagation among Japanese macaques. *Primates*, **2**, 43–55.

Kay, R. F. (1985). Dental evidence for the diet of *Australopithecus*. *Annual Review of Anthropology*, **14**, 315–41.

Kelch, R., Jenner, M., Weinstein, R., Kaplan, S. & Grumbach, M. (1972). Estradiol and testosterone secretion by human, simian and canine testes, in males with hypogonadism and in male pseudo hermaphrodites with the feminizing testes syndrome. *Journal of Clinical Investigation*, **51**, 824–30.

Kelch, R., Hopwood, N. J., Sauder, S. & Marshall, J. C. (1985). Evidence for decreased secretion of gonadotropin-releasing hormone in pubertal boys during short-term testosterone treatent. *Pediatric Research*, **19**, 113–17.

Keyes, R. (1979). The height report. *Esquire*, November, 31–43.

Kimura, K. (1984). Studies on growth and development in Japan. *Yearbook of Physical Anthropology*, **27**, 179–214.

Kimura, K. & Kitano, S. (1959). Growth of the Japanese physiques in four successive decades before World War II. *Zinruigaku Zassi*, **67**, 37–46.

King, M. C. & Wilson, A. C. (1975). Evolution at two levels: molecular similarities and differences between humans and chimpanzees. *Science*, **188**, 107–16.

Kobyliansky, E. & Arensburg, B. (1974). Changes in morphology of human populations due to migrations and selection. *Annals of Human Biology*, **4**, 57–71.

Kondo, S. & Eto, M. (1975). Physical growth studies on Japanese–American children in comparison with native Japanese. In *Comparative Studies of Human Adaptability of Japanese, Caucasians, and Japanese-Americans*, ed. S. M. Horvath, S. Kondo, H. Matsui & H. Yoshimena. pp. 13–45. Tokyo: Japanese International Biological Program.

Konner, M. (1976). Maternal care, infant behavior and development among the !Kung. In *Kalahari Hunter-Gatherers*, ed. R. B. Lee & I. DeVore, pp. 218–45. Cambridge: Harvard University Press.

Krogman, W. M. (1970). Growth of the head, face, trunk, and limbs in Philadelphia white and Negro children of elementary and high school age. *Monographs of the Society for Research in Child Development*, **20**, 1–91.

Kummer, B. (1953). Untersuchungen über die ontogenetische Entwicklung des menschlichen Schadelbasiswinkels. *Zeitschrift für Morphologie und Anthropologie*, **43**, 331–60.

Laird, A. K. (1967). Evolution of the human growth curve. *Growth*, **31**, 345–55.

Laitman, J. T. & Heimbuch, R. C. (1982). The basicranium of Plio-Pleistocene hominids as an indicator of their upper respiratory systems. *American Journal Physical Anthropology*, **59**, 323–43.

Lancaster, J. B. (in press). Evolutionary perspectives on sex differences in the higher primates. In *Gender and the Life Course*, ed. A. S. Rossi. New York: Aldine.

Lancaster, J. B. & Lancaster, C. S. (1983). Parental investment: The hominid adaptation. In *How Humans Adapt*, ed. D. J. Ortner, pp. 33–65. Washington DC: Smithsonian Institution Press.

Lancaster, J. B. & Whitten, P. (1985). Sharing in human evolution. In *Anthropology: Contemporary Perspectives*, 4th edn, ed. D. E. K. Hunter & P. Whitten, pp. 45–49. Boston: Little, Brown.

Largo, R. H., Gasser, Th., Prader, A., Stutzle, W. & Huber, P. J. (1978). Analysis of the adolescent growth spurt using smoothing spline functions. *Annals of Human Biology*, **5**, 421–34.

Laron, Z., Arad, J., Gurewitz, R., Grunebaum, M. & Dickerman, Z. (1980). Age at first conscious ejaculation – a milestone in male puberty. *Helvatica Paediatrica Acta*, **35**, 13–20.

Lasker, G. W. (1952). Environmental growth factors and selective migration. *Human Biology*, **24**, 262–89.

Lasker, G. W. (1954). The question of physical selection of Mexican migrants to the United States of America. *Human Biology*, **26**, 52–8.

Lasker, G. W. (1969). Human biological adaptability. *Science*, **166**, 1480–6.

Lee, M. M. C., Chang, K. S. F. & Chan, M. M. C. (1963). Sexual maturation of Chinese girls in Hong Kong. *Pediatrics*, **32**, 389–98.

Leighton, G. & Clark, M. L. (1929). Milk consumption and the growth of schoolchildren. *Lancet*, **1**, 40–3.

Lenko, H. L., Laing, U., Aubert, M. L., Paunier, L. & Sizonenko, P. C. (1981). Hormonal changes in puberty, VII: Lack of variation of daytime melatonin. *Journal of Clinical Endocrinology and Metabolism*, **54**, 1056–8.

Lewin, R. (1985). Dental humans, infant apes. *Science*, **230**, 795.

Lichty, J. A., Ting, R. Y., Bruns, P. & Dyar, E. (1957). Studies of babies born at high altitude. I. Relation of altitude to birth-weights. *American Journal of Disease in Childhood*, **93**, 666–9.

Lieberman, P., Crelin, E. S. & Klatt, D. H. (1972). Phonetic ability and related anatomy of the newborn and adult human, Neanderthal man, and the chimpanzee. *American Anthropologist*, **74**, 287–307.

Lindgren, G. (1976). Height, weight, and menarche in Swedish urban school children in relation to socio-economic and regional factors. *Annals of Human Biology*, **3**, 501–28.

Lindgren, G. (1978). Growth of schoolchildren with early, average, and late ages of peak height velocity. *Annals of Human Biology*, **5**, 253–67.

Little, M. A., Galvin, K. & Mugambi, M. (1983). Cross-sectional growth of nomadic Turkana pastoralists. *Human Biology*, **55**, 811–30.

Livi, R. (1896). *Antropometrica Militare*. Rome, cited in Boas (1922).

Lorenz, K. (1971). Part and parcel in animal and human societies: a methodological discussion. In *Studies in Animal and Human Behavior*, Vol. II (translated by Robert Martin), pp. 115–95. Cambridge: Harvard University Press.

Lourie, J. A., Taufa, T., Cattani, J. & Anderson, W. (1986). The Ok Tedi Health and Nutrition Project, Papua New Guinea: physique, growth and nutritional status of the Wopkaimin of the Star Mountains. *Annals of Human Biology*, **13**, 517–36.

Lovejoy, A. O. (1936). *The Great Chain of Being*. Cambridge: Harvard University Press.

Lovejoy, C. O. (1973). The gait of australopithecines. *Yearbook of Physical Anthropology*, **17**, 147–61.

Lovejoy, C. O., Meindl, R. S., Pryzbeck, T. R., Barton, T. S., Heiple, K. G. & Kotting, D. (1977). Paleodemography of the Libben site, Ottawa County, Ohio. *Science*, **198**, 291–3.

Lovejoy, O. (1981). The origin of man. *Science*, **211**, 341–50.

Low, W. D., Kung, L. S. & Leong, J. C. Y. (1982). Secular trend in the sexual maturation of Chinese girls. *Human Biology*, **54**, 539–51.

Lowe, C. U., Forbes, G., Garn, S., Owen, G. M., Smith, N. J., Weil, W. B. Jr. & Nichaman, M. Z. (1975). Reflections of dietary studies with children in the Ten-State Nutrition Survey of 1968–1970. *Pediatrics*, **56**(2), 320–6.

Lowery, G. H. (1986). *Growth and Development of Children*, 8th edn. Chicago: Yearbook Medical Publishers.

Luft, U. C. (1972). Principles of adaptation to altitude. In *Physiological Adaptations: Desert and Mountain*, ed. M. K. Yousef, S. M. Horvath & R. W. Ballard, pp. 143–56. New York: Academic Press.

MacBeth, H. M. (1984). The study of biological selectivity in migrants. In *Migration and Mobility*, ed. A. J. Boyce, pp. 195–207. London: Taylor & Francis.

Magner, J. A., Rogol, A. D. & Gorden, P. (1984). Reversible growth hormone deficiency and delayed puberty triggered by a stressful experience in a young adult. *American Journal of Medicine*, **76**, 737–42.

Malcolm, L. A. (1970). Growth and development in the Bundi children of the New Guinea highlands. *Human Biology*, **42**, 293–328.

Malina, R. M. (1966). Patterns of development in skinfolds of Negro and white Philadelphia children. *Human Biology*, **38**, 89–103.

Malina, R. M. (1979). Secular changes in size and maturity: causes and effects. In Secular Trends in Human Growth, maturation, and development, ed. A. F. Roche, *Monographs of the Society for Research in Child Development*, **44**, 59–102.

Malina, R. M., Harper, A. B., Avent, H. H. & Campbell (1973). Age at menarche in athletes and non-athletes. *Medicine and Science in Sports*, **5**, 11–13.

Malina, R. M., Selby, H. A., Bushang, P. H. & Aronson, W. L. (1980). Growth status of schoolchildren in a rural Zapotec community in the Valley of Oaxaca, Mexico, in 1968 and 1978. *Annals of Human Biology*, **7**, 367–74.

Malina, R. M., Himes, J. H., Stepick, C. D., Lopez, F. G. & Buschang, P. H. (1981). Growth of rural and urban children in the Valley of Oaxaca, Mexico. *American Journal of Physical Anthropology*, **55**, 269–80.

Malina, R. M., Bushang, P. H., Aronson, W. L. & Selby, H. (1982a). Childhood growth status of eventual migrants and sedentes in a rural Zapotec community in the valley of Oaxaca, Mexico. *Human Biology*, **54**, 709–16.

Malina, R. M., Mueller, W. H., Bouchard, C., Shoup, R. F. & Lariviere, G. (1982b). Fatness and fat patterning among athletes at the Montreal Olympic Games, 1976. *Medicine and Science in Sports and Exercise*, **14**, 445–52.

Malina, R. M., Little, B. B., Stern, M. P., Gaskill, S. P. & Hazuda, H. P. (1983). Ethnic and social class differences in selected anthropometric characteristics of Mexican American and Anglo adults: The San Antonio heart study. *Human Biology*, **55**, 867–83.

Malina, R. M., Little, B. B., Buschang, P. H., DeMoss, J. & Selby, H. A. (1985). Socioeconomic variation in the growth status of children in a subsistence agricultural community. *American Journal of Physical Anthropology*, **68**, 385–91.

Mann, A. E. (1968). *The Paleodemography of Australopithecus*. Dissertation. Ann. Arbor: University Microfilms.

Mann, A. E. (1972). Hominid and cultural origins. *Man*, **7**, 379–86.

Mann, A. E. (1975). Some paleodemographic aspects of the South African australopithecines. *University of Pennsylvania Publications in Anthropology*, No. 1. Philadelphia.

Mann, G. V., Roels, O. A., Price, D. L. & Merrill, J. M. (1962). A survey of the health status, serum lipids and diet of Pygmies in Congo. *Journal of Chronic Diseases*, **15**, 341–71.

Markowitz, S. D. (1955). Retardation in growth of children in Europe and Asia during World War II. *Human Biology*, **27**, 258–73.

Marshall, J. C. & Kelch, R. P. (1986). Neuroendocrine regulation of reproduction. The critical role of pulsatile GnRH secretion and implications for therapy. *New England Journal of Medicine*, **315**, 1459–67.

Marshall, W. A. (1975). The relationship of variations in children's growth rates to season climatic variation. *Annals of Human Biology*, **2**, 243–50.

Marshall, W. A. (1978). Puberty. In *Human Growth*, Vol. 2, ed. F. Faulkner & J. M. Tanner, pp. 141–81. New York: Plenum.

Marshall, W. A. & Swan, A. V. N. (1971). Seasonal variation in growth rates of normal and blind children. *Human Biology*, **43**, 502–16.

Marshall, W. A. & Tanner, J. M. (1969). Variation in the pattern of pubertal changes in girls. *Archives of the Diseases of Childhood*, **44**, 291–303.

Marshall, W. A. & Tanner, J. M. (1970). Variation in the pattern of pubertal changes in boys. *Archives of the Diseases of Childhood*, **45**, 13–23.

Marshall, W. A. & Tanner, J. M. (1986). Puberty. In *Human Growth*, Vol. 2, 2nd edn, ed. F. Faulkner & J. M. Tanner, pp. 171–209. New York: Plenum.

Martin, D. E., Swenson, R. B. & Colins, D. C. (1977). Correlation of serum testosterone levels with age in male chimpanzees. *Steroids*, **29**, 471–81.

Martin, R. D. (1968). Reproduction and ontogeny in tree shrews (*Tupaia belangeri*) with reference to their general behavior and taxonomic relationships. *Zeitschrift für Tierpsychologie*, **25**, 409–95, 505–32.

Martin, W. J. (1949). The physique of young adult males. *Medical Research Council Memorandum*, No. 20. London: HMSO.

Martorell, R., Yarbrough, C., Lechtig, A., Delgado, H. & Klein (1976). Upper arm anthropometric indicators of nutritional status. *American Journal of Clinical Nutrition*, **29**, 46–53.

Martorell, R., Yarbrough, C., Lechtig, A., Delgado, H. & Klein, R. E. (1977). Genetic–environmental interactions in physical growth. *Acta Paediatrica Scandinavia*, **66**, 579–84.

Martorell, R., Delgado, H. L., Valverde, V. & Klein, R. E. (1981). Maternal stature, fertility and infant mortality. *Human Biology*, **53**, 303–12.

Marubini, E. (1978). The fitting of longitudinal growth data of man. In *Auxology: Human Growth in Health and Disorder*, ed. L. Gedda & P. Parisi, pp. 123–32. London: Academic Press.

Marubini, E. & Milani, S. (1986). Approaches to the analysis of longitudinal data. In *Human Growth*, Vol. 3, 2nd edn, ed. F. Falkner & J. M. Tanner, pp. 79–109. New York: Plenum.

Marubini, E., Resele, L. F., Tanner, J. M. & Whitehouse, R. H. (1972). The fit of Gompertz and logistic data during adolescence on height, sitting height and biacromial diameter in boys and girls of the Harpenden Growth Study. *Human Biology*, **44**, 511–24.

Mascie-Taylor, C. G. N. (1984). The interaction between geographical and social mobility. In *Migration and Mobility*, ed. A. J. Boyce, pp. 161–78. London: Taylor & Francis.

Mascie-Taylor, C. G. N. & Boldsen, J. L. (1985). Regional and social analysis of height variation in a contemporary British sample. *Annals of Human Biology*, **12**, 315–24.

Mascie-Taylor, C. G. N. & Lasker, G. W., eds. (1988). *Biological Aspects of Human Migration*. Cambridge: Cambridge University Press.

Masse, G. & Hunt, E. E. Jr. (1963). Skeletal maturation in the hand and wrist of West African children. *Human Biology*, **35**, 3–25.

Matsumoto, K. (1982). Secular acceleration of growth in height in Japanese and its social background. *Annals of Human Biology*, **9**, 399–410.

Mazess, R. B. & Mathisen, R. W. (1982). Lack of unusual longevity in Vilcabamba, Ecuador. *Human Biology*, **54**, 517–24.

McCullough, J. M. (1982). Secular trend for stature in adult male Yucatec Maya to 1968. *American Journal of Physical Anthropology*, **58**, 221–5.

McGarvey, S. T. (1984). Subcutaneous fat distribution and blood pressure of Samoans. (Abstract) *American Journal of Physical Anthropology*, **63**, 192.

McKinley, K. R. (1971). Survivorship in gracile and robust australopithecines: a demographic comparison and a proposed birth model. *American Journal of Physical Anthropology*, **34**, 417–26.

Medawar, P. B. (1945). Size, shape and age. In *Essays on Growth and Form*, ed. W. E. LeGros Clark & P. B. Medawar, pp. 157–87. Oxford: Clarendon Press.

Meire, H. B. (1986). Ultrasound measurement of fetal growth. In *Human Growth*, Vol. 1, 2nd edn, ed. F. Falkner & J. M. Tanner, pp. 275–90. New York: Plenum.

Mellits, E. D. & Cheek, D. B. (1968). Growth and body water. In *Human Growth*, ed. D. B. Cheek, pp. 135–49. Philadelphia: Lee & Febiger.

Mendez, J. & Behrhorst, C. (1963). The anthropometric characteristics of Indians and urban Guatemalans. *Human Biology*, **36**, 457–69.

Meredith, H. V. (1935). The rhythm of physical growth. *University of Iowa Studies in Child Welfare*, **11**, 124.

Meredith, H. V. (1979). Comparative findings on body size of children and youths living at urban centers and in rural areas. *Growth*, **43**, 95–104.

Meredith, H. V. (1982). An addendum on presence or absence of a mid-childhood spurt in somatic dimensions. *Annals of Human Biology*, **8**, 473–6.

Merimee, T. J., Rimoin, D. L., Rabinowitz, D., Cavalli-Sforza, L. L. & McKusick, V. A. (1968). Metabolic studies in the African Pygmy. *Transactions of the Association of American Physicians*, **81**, 221–320

Merimee, T. J., Zapf, J., Froesch, E. R. (1981). Dwarfism in the pygmy, an isolated deficiency of insulin-like growth factor I. *New England Journal of Medicine*, **305**, 965–8.

Milton, K., (1983). Morphometric features as tribal predictors in North-Western Amazonia. *Annals of Human Biology*, **10**, 435–40.

Mitton, J. B. (1975). Fertility differentials in modern societies resulting in normalizing selection for height. *Human Biology*, **47**, 189–200.

Molinari, L., Largo, R. H. & Prader, A. (1980). Analysis of the growth spurt at age seven (mid-growth spurt). *Helvetica Paediatrica Acta*, **35**, 325–34.

Morikawa, M., Nixon, T. & Green, H. (1982). Growth hormone and the adipose conversion of 3T3 cells. *Cell*, **29**, 783–9.

Mossman, H. W. (1937). Comparative morphogenesis of the fetal membranes and accessory uterine structures. *Contributions to Embryology*, **26**, 129–246.

Mueller, W. H. (1977). Sibling correlations in growth and morphology in a rural Columbian population. *Annals of Human Biology*, **4**, 133–42.

Mueller, W. H. (1979). Fertility and physique in a malnourished population. *Human Biology*, **51**, 153–66.

Mueller, W. H. (1982). The changes with age of the anatomical distribution of fat. *Social Science and Medicine*, **16**, 191–6.

Mueller, W. H. & Pollitt, E. (1982). The Bacon Chow study: effects of nutritional supplementation on sibling-sibling anthropometric correlations. *Human Biology*, **54**, 455–68.

Mueller, W. H. & Pollitt, E. (1983). The Bacon Chow Study: genetic analysis of physical growth in assessment of energy–protein malnutrition. *American Journal of Physical Anthropology*, **62**, 11–17.

Mueller, W. H., Schull, V. N., Schull, W. J. (1978). A multinational Andean genetic and health program: growth and development in an hypoxic environment. *Annals of Human Biology*, **5**, 329–52.

Mueller, W. H., Murillo, F., Palamino, H., Badzioch, M., Chakraborty, R., Fuerst, P. & Schull, W. J. (1980). The Aymara of Western Bolivia: V. Growth and development in an hypoxic environment. *Human Biology*, **52**, 529–46.

Mueller, W. H., Joos, S. K., Janis, C. L., Zavalita, A. N., Eicher, J. & Schull, W. J. (1984) Diabetes alert study: growth fatness, and fat patterning, adolescence through adulthood, in Mexican–Americans. *American Journal of Physical Anthropology*, **64**, 389–99.

Neel, J. V. & Weiss, K. (1975). The genetic structure of a tribal population, the Yanomamo Indians. Biodemographic studies XII. *American Journal of Physical Anthropology*, **42**, 25–52.

Nellhaus, G. (1986). Head circumference from birth to eighteen years. *Pediatrics*, **41**, 106.

Newell–Morris, L. & Fahrenbach, C. F. (1985). Practical and evolutionary considerations for use of the non-human primate model in pre-natal research. In *Non-human Primate Models for Human Growth and Development*, ed. E. Watts, pp. 9–40. New York: Alan R. Liss.

Newman, M. T. (1953). The application of ecological rules to racial anthropology of the aboriginal new-world. *American Anthropologist*, **55**, 311–27.

Newman, R. W. (1970). Why man is such a sweaty and thirsty naked animal: a speculative review. *Human Biology*, **42**, 12–27.

Neyzi, O., Alp, H. & Orhon, A. (1975a). Sexual maturation in Turkish girls. *Annals of Human Biology*, **2**, 49–59.

Neyzi, O., Alp, H., Yalcindag, A., Yakacikli, S. & Orhon, A. (1975b). Sexual maturation in Turkish boys. *Annals of Human Biology*, **2**, 251–9.

Nicolson, A. B. & Hanley, C. (1953). Indices of physiological maturity: derivation and interrelationships. *Child Development*, **24**, 3–38.

Nilsson, A., Isgaard, J., Lindahl, A., Dahlstrom, A., Skottner, A. & Isaksson, O. G. P. (1986). Regulation by growth hormone of number of chondrocytes containing IGF-I in rat growth plate. *Science*, 233, 571–4.

Nylin, G. (1929). Periodical variation in growth, standard metabolism and oxygen capacity of the blood in children. *Acta Medica Scandinavia*, 31, 1–207.

Nixon, T. & Green, H. (1984). Contribution of growth hormone to the adipogenic activity of serum. *Endocrinology*, 114, 527–32.

Onat, T. & Ertem, B. (1974). Adolescent female height velocity: relationships to body measurements, sexual and skeletal maturity. *Human Biology*, 46, 199–217

O'Rahilly, R. & Muller, F. (1986). Human growth during the embryonic period proper. In *Human Growth*, Vol. 1, 2nd edn, ed. F. Falkner & J. M. Tanner, pp. 245–53. New York: Plenum.

Orlosky, F. J. (1982). Adolescent midfacial growth in Macaca nemestrina and Papio cynocephalus. *Human Biology*, 54, 23–9.

Orr, J. B. (1928). Milk consumption and the growth of schoolchildren. *Lancet*, 1, 202–3.

Panek, S. & Piasecki, M. (1971). Nowa Huta: integration of the population in the light of anthropometric data. *Materialyi I Prace Anthropologiczne*, 80, 1–249 (in Polish with English summary).

Parisi, P. & de Martino, V. (1980). Psychosocial factors in human growth. In *Human Physical Growth and Maturation: Methodologies and Factors*, ed. F. E. Johnston, A. F. Roche & C. Susanne, pp. 339–56. New York: Plenum.

Parnell, R. W. (1954). The physique of Oxford undergraduates. *Journal of Hygiene*, 52, 396–78.

Pawson, I. G. (1977). Growth characteristics of populations of Tibetan origin in Nepal. *American Journal of Physical Anthropology*, 47, 473–82.

Pawson, I. G. & Janes, C. (1981). Massive obesity in a migrant Samoan population. *American Journal of Public Health*, 71, 508–13.

Payne, P. R. & Waterlow, J. C. (1971). Relative energy requirements for maintenance, growth and physical activity. *Lancet*, 2(1), 210–11.

Perieira, M. E. & Altman, J. (1985). Development of social behavior in free-living nonhuman primates. In *Nonhuman Primate Models for Human Growth and Development*, ed. E. S. Watts, pp. 217–309. New York: Alan R. Liss.

Petersen, A. C. & Taylor, B. (1980). The biological approach to adolescence: biological change and psychological adaptation. In *Handbook of Adolescent Psychology*, ed. J. Adelson, pp. 117–55. New York: Wiley.

Piaget, J. (1954). *The Construction of Reality in the Child*. New York: Basic Books.

Piaget, J. & Inhelder, B. (1969). *The Psychology of the Child*. New York: Basic Books.

Pilbeam, D. (1979). Recent finds and interpretations of Miocene hominoids. *Annual Review of Anthropology*, 8, 833–52.

Pilbeam, D. (1984). The descent of hominoids and hominids. *Scientific American*, 250, 84–96.

Pimstone, B. L., Barbezat, G., Hansen, J. D. & Murray, P. (1968). Studies on growth hormone secretion in protein–calorie malnutrition. *American Journal of Clinical Nutrition*, 21, 482–7.

Piscopo, J. (1962). Skinfold and other anthropometrical measures of pre-adolescant boys from three ethnic groups. *Research Quarterly*, 33, 255–62.

Plotsky, P. M. & Vale, W. (1985). Patterns of growth hormone-releasing factor and somatostatin secretion into the hypophysial-portal circulation of the rat. *Science*, **230**, 461–3.

Potts, R. (1984). Home bases and early hominids. *American Scientist*, **72**, 338–47.

Powell, G. F., Brasel, J. A. & Blizzard, R. M. (1967a). Emotional deprivation and growth retardation simulating idiopathic hypopituitarism. I. Clinical evaluation of the syndrome. *New England Journal of Medicine*, **276**, 1271–8.

Powell, G. F., Raiti, S. & Blizzard, R. M. (1967b). Emotional deprivation and growth retardation simulating idiopathic hypopituitarism. II. Endocrine evaluation of the syndrome. *New England Journal of Medicine*, **276**, 1279–83.

Prader, A. (1984). Biomedical and endocrinological aspects of normal growth and development. In *Human Growth and Development*, ed. J. Borms, R. Hauspie, A. Sand, C. Susanne & M. Hebbelinck, pp. 1–22. New York: Plenum.

Prader, A., Tanner, J. M. & Von Harnack, G. A. (1963). Catch-up growth following illness or starvation. *Journal of Paediatrics*, **62**, 645–59.

Preece, M. A. (1986). Prepubertal and pubertal endocrinology. In *Human Growth*, Vol. 2, 2nd edn, ed. F. Faulkner & J. M. Tanner, pp. 211–24. New York: Plenum.

Preece, M. A. & Baines, M. J. (1978). A new family of mathematical models describing the human growth curve. *Annals of Human Biology*, **5**, 1–24.

Preece, M. A. & Holder, A. T. (1982). The somatomedins: A family of serum growth factors. In *Recent Advances in Endocrinology and Metabolism*, Vol. 2, ed. J. L. H. O'Riordan, pp. 47–72. Edinburgh: Churchill Livingstone.

Preece, M. A., Cameron, N., Donnall, M. C., Dunger, D. B., Holder, A. T., Baines-Preece, J., Seth, J., Sharp, G. & Taylor, A. M. (1984). The endocrinology of male puberty. In *Human Growth and Development*, ed. J. Borms, R. Hauspie, A. Sand, C. Susanne & M. Hebbelinck, 23–37. New York: Plenum.

Quo, S.-K. (1955). Mathematical analysis of the growth of man, with special reference to Formosans. *Human Biology*, **25**, 333–58.

Ralston, A. (1986). Discrete mathematics: the new mathematics of science. *American Scientist*, **74**, 611–18.

Ramirez, M. E. & Mueller, W. H. (1980). The development of obesity and fat patterning in Tokelau children. *Human Biology*, **52**, 675–88.

Rappaport, R. (1984). Growth hormone secretion in children of short stature. In *Human Growth*, ed. J. Borms, R. Hauspie, A. Sand, C. Susanne & M. Hebbelinck, pp. 39–48. New York: Plenum.

Rasmusen, H. (1974). Parathyroid hormone, calcitonin and the caliciferols. In *Textbook of Endocrinology*, ed. B. H. Williams, pp. 660–773. Philadelphia: Saunders.

Ratcliffe, S. G. (1976). The development of children with sex chromosome abnormalities. *Proceedings of the Royal Society of Medicine*, **69**, 189–91.

Ratcliffe, S. G. (1981). The effect of chromosome abnormalities on human growth. *British Medical Bulletin*, **37**, 291–5.

Reynolds, E. L. & Wines, J. V. (1948). Individual differences in physical changes associated with adolescence in girls. *American Journal of Disease of Children*, **75**, 329–50.

Richardson, D. W. & Short, R. V. (1978). Time of onset of sperm production in boys. *Journal of Biosocial Science* (Supplement) **5**, 15–24.

Rimoin, D. L., Merimee, T. J., Rabinowitz, D., Cavalli-Sforza, L. L. & McKusick, V. A. (1968). Genetic aspects of clinical endocrinology. *Recent Progress in Hormone Research*, **24**, 365–467.

Roberts, D. F. (1953). Body weight, race and climate. *American Journal of Physical Anthropology*, **11**, 533–58.

Roberts, D. F., Billewicz, W. Z. & McGregor, I. A. (1978). Heritability of stature in a west African population. *Annals of Human Genetics*, **42**, 15–24.

Robertson, T. B. (1908). On the normal rate of growth of an individual, and its biochemical significance. *Archiv für Entwicklungs Mechanik den Organismen*, **25**, 581–614.

Robson, E. B. (1978). The genetics of birth weight. In *Human Growth*, Vol. 1, ed. F. Faulkner & J. M. Tanner, pp. 285–97. New York: Plenum.

Robson, J. R. K., Bazin, M. & Soderstrom, B. S. (1971). Ethnic differences in skin-fold thickness. *American Journal of Clinical Nutrition*, **24**, 864–8.

Roche, A. F. & Davila, G. H. (1972). Late adolescent growth in stature. *Pediatrics*, **50**, 874–80.

Roche, A. F., Wainer, H. & Thissen, D. (1975a). *Skeletal Maturity. The Knee Joint as a Biological Indicator*. New York: Plenum.

Roche, A. F., Wainer, H. & Thissen, D. (1975b). Predicting adult stature for individuals. *Monographica Paediatrica*, **3**, 41–96.

Roede, M. J. (1985). The privilege of growing. *Acta Medica Auxologica*, **17**, 217–26.

Rogers, A. & Williamson, J. C. (1982). Migration, urbanization, and third world development: and overview. *Economic Development and Cultural Change*, **30**, 463–82.

Romer, A. S. (1966). *Vertebrate Paleontology.* Chicago: University of Chicago Press.

Rona, R. J. & Chinn, S. (1986). National study of health and growth: social and biological factors associated with height of children from ethnic groups living in England. *Annals of Human Biology*, **13**, 453–71.

Rosenfeld, R. L., Furlanetto, R. & Bock, D. (1983). Relationship of somatomedin-C concentrations to pubertal changes. *Journal of Pediatrics*, **103**, 723–8.

Rosenwaike, I. & Preston, S. H. (1984). Age overstatement and Puerto Rican longevity. *Human Biology*, **56**, 503–25.

Ruble, D. N. & Brooks-Gunn, J. (1982). The experience of menarche. *Child Development*, **53**, 1557–66.

Russell, M. (1976). Parent–child and sibling–sibling correlations of height and weight in a rural Guatemalan population of preschool children. *Human Biology*, **48**, 501–15.

Sabharwal, K. P., Morales, S. & Mendez, J. (1966). Body measurements and creatinine excretion among upper and lower socioeconomic groups of girls in Guatemala. *Human Biology*, **38**, 131–40.

Sacher, G. A. (1975). Maturation and longevity in relation to cranial capacity in hominid evolution. In *Primate Functional Morphology and Evolution*, ed. R. Tuttle, pp. 417–41. The Hague: Mouton.

Saenger, P., Levine, L. S., Wiedemann, E., Schwartz, E., Korth-Schutz, S., Paretra, J., Heinig, B. & New, M. I. (1977). Somatomedin and growth hormone in psychosocial dwarfism. *Padiatrie und Padologie Supplementum*, **5**, 1–12.

Sanger, R. Tippett, P. & Gavin, J. (1971). Xg groups and sex abnormalities in people of Northern European ancestry. *Journal of Medical Genetics*, **8**, 417–26.

Satyanarayana, K., Naidu, A. N. & Rao, B. S. N. (1980). Adolescent growth spurt among rural Indian boys in relation to their nutritional status in early childhood. *Annals of Human Biology*, **7**, 359–65.

Scammon, R. E. (1927). The first seriation study of human growth. *American Journal of Physical Anthropology*, **10**, 329–36.

Scammon, R. E. (1930). The measurement of the body in childhood. In *The Measurement of Man*, ed. J. A. Harris, C. M. Jackson, D. G. Paterson & R. E. Scammon, pp. 173–215. Minneapolis: University of Minnesota Press.

Scammon, R. E. & Calkins, L. A. (1929). *The Development and Growth of the External Dimensions of the Human Body in the Fetal Period*. Minneapolis: University of Minnesota Press.

Schally, A. V., Kastin, A. J. & Arimura, A. (1977). Hypothalmic hormones: the link between brain and body. *American Scientist*, **65**, 712–19.

Schell, L. M. & Hodges, D. C. (1985). Variation in size at birth and cigarette smoking during pregnancy. *American Journal of Physical Anthropology*, **68**, 549–54.

Schofield, M. (1965). *The Sexual Behavior of Young People*. London: Longman's Green.

Schreider, E. (1964a). Ecological rules, body-heat regulation, and human evolution. *Evolution*, **18**, 1–9.

Schreider, E. (1964b). Recherches sur la stratification sociale des caractères biologiques. *Biotypologie*, **26**, 105–35.

Schultz, A. H. (1969). *The Life of Primates*. New York: Universe Books.

Schwartz, J., Brumbaugh, R. C. & Chiu, M. (under review). Short stature, growth hormone, insulin-like growth factors, and serum proteins in the Mountain Ok people of Papua New Guinea.

Scott, E. C. & Bajema, C. J. (1982). Height, weight and fertility among the participants of the Third Harvard Growth Study. *Human Biology*, **54**, 501–16.

Scott, E. C. & Johnston, F. E. (1982). Critical fat, menarche, and the maintenance of menstrual cycles. *Journal of Adolescent Health Care*, **2**, 249–60.

Scott, E. M., Illsley, I. P. & Thomson, A. M. (1956). A psychological investigation of primigravidae. Maternal social class, age, physique and intelligence. *Journal of Obstetrics and Gynaecology of the British Empire*, **63**, 338–43.

Scott, J. P. (1967). Comparative psychology and ethnology. *Annual Review of Psychology*, **18**, 65–86.

Service, E. R. (1978). The Arunta of Australia. In *Profiles in Ethnology*, ed. E. R. Service, pp. 13–34. New York: Harper & Row.

Shapiro, H. L. (1939). *Migration and Environment*. Oxford: Oxford University Press.

Shapiro, S. & Unger, J. (1965). *Relation of weight at birth to cause of death and age at death in the neonatal period: United States, early 1950*. Public Health Service Pub. No. 1000–Series 21–No. 6. Washington DC: US Government Printing Office.

Shock, N. W. (1966). Physiological growth. In *Human Development*, ed. F. Falkner, pp. 150–77. Philadelphia: Saunders.

Shohoji, T. & Sasaki, H. (1984). The growth process of the stature of Japanese: Growth from early childhood. *Acta Medica Auxologica*, **16**, 101–11.

Short, R. V. (1976). The evolution of human reproduction. *Proceedings, Royal Society*, Series B, **195**, 3–24.

Shuttleworth, F. K. (1937). Sexual maturation and the physical growth of girls age six to nineteen. *Monographs of the Society for Research in Child Development*, **2**, No. 5.

Shuttleworth, F. K. (1939). The physical and mental growth of girls and boys age six to nineteen in relation to age at maximum growth. *Monographs of the Society for Research in Child Development*, **4**, No. 3.

Sills, R. H. (1978). Failure to thrive: the role of clinical and laboratory evaluation. *American Journal of Diseases of Children*, **132**, 967–9.

Silman, R. E., Leone, R. M., Hooper, R. J. L. & Preece, M. A. (1979). Melatonin, the pineal gland and human puberty. *Nature*, **282**, 301–2.

Simmons, K. & Greulich, W. W. (1943). Menarcheal age and the height, weight and skeletal age of girls, age 7 to 17 years. *Journal of Pediatrics*, **22**, 518–48.

Simons, E. L. & Pilbeam, D. R. (1972). Hominoid paleoprimatology. In *The Functional and Evolutionary Biology of Primates*, ed. R. Tuttle, pp. 36–62. Chicago: Aldine.

Singer, C. (1959). *A Short History of Scientific Ideas to 1900*. London: Oxford University Press.

Sirianni, J. E., VanNess, A. L. & Swindler, D. R. (1982). Growth of the mandible in adolescent pigtailed macaques (*Macaca nemestrina*). *Human Biology*, **54**, 31–44.

Sizonenko, P. C. & Aubert, M. L. (1986). Pre- and perinatal endocrinology. In *Human Growth*, Vol. 1, 2nd edn, ed. F. Falkner & J. M. Tanner, pp. 339–76. New York: Plenum.

Sizonenko, P. C. & Paunier, L. (1982). Failure of DHEA-oenenthate to promote growth. *Pediatric Research*, **16**, 888 (abstract).

Smith, B. H. (1986). Dental development in *Australopithecus* and early *Homo*. *Nature*, **323**, 327–30.

Smith, M. T. (1984). The effects of migration on sampling in genetical surveys. In *Migration and Mobility*, ed. A. J. Boyce, pp. 97–110. London: Taylor & Francis.

Snow, M. H. L. (1986). Control of embryonic growth rate and fetal size in mammals. In *Human Growth*, Vol. 1, 2nd edn, ed. F. Falkner & J. M. Tanner, pp. 67–82. New York: Plenum.

Spitz, R. A. (1945). Hospitalism: An inquiry into the genesis of psychiatric conditions in early childhood. *Psychoanalytic Study of the Children*, **1**, 53–74.

Stamp, T. C. B. & Round, J. M. (1974). Seasonal changes in human plasma levels of 25-hydroxy vitamin D. *Nature*, **247**, 563–5.

Starck, D. & Kummer, B. (1962). Zur Ontogenese des Schimpansenschadels. *Anthropologischer Anzieger*, **25**, 204–15.

Steegmann, A. T. Jr. (1985). 18th century British military stature: growth cessation, selective recruiting, secular trends, nutrition at birth, cold and occupation. *Human Biology*, **57**, 77–95.

Stein, Z., Susser, M., Saenger, G. & Marolla, F. (1975). *Famine and Human Development. The Dutch Hunger Winter of 1944–1945*. London: Oxford University Press.

Stern, M. P., Haskell, W. L., Wood, P. D. S., Osann, K. E., King, A. B. & Farquhar, J. W. (1975). Affluence and cardiovascular risk factors in Mexican-Americans and other whites in three northern California communities. *Journal of Chronic Diseases*, **28**, 623–36.

Stini, W. A. (1975). Adaptive strategies of human populations under nutritional stress. In *Biosocial Interrelations in Population Adaptation*, ed. E. S. Watts, F. E. Johnston & G. W. Lasker, pp. 19–41. The Hague: Mouton.

Stinson, S. (1982). The effect of high altitude on the growth of children of high socioeconomic status in Bolivia. *American Journal of Physical Anthropology*, **59**, 61–71.

Stratz, C. H. (1909). Wachstum und Proportionen des Menschen vor und nach der Geburt. *Archiv für Anthropologie*, **8**, 287–97.

Stutzle, W., Gasser, Th., Molinari, L., Largo, R. H., Prader, A. & Huber, P. S. (1980). Shape-invariant modelling of human growth. *Annals of Human Biology*, **7**, 507–28.

Sullivan, P. G. (1986). Skull, jaw and teeth growth patterns. In *Human Growth*, Vol. 2, ed. F. Falkner & J. M. Tanner, pp. 243–68. New York: Plenum.

Susanne, C. (1975). Genetic and environmental influences on morphological characteristics. *Annals of Human Biology*, **2**, 279–87.

Susanne, C. (1980). Interrelations between some social and familial factors and stature and weight of young Belgian male adults. *Human Biology*, **52**, 701–9.

Szathmary, E. J. E. & Holt, N. (1983). Hypoglycemia in Dogrib Indians of the Northwest Territories, Canada: association with age and a centripetal distribution of body fat. *Human Biology*, **55**, 493–515.

Taffel, S. (1980). *Factors associated with low birth weight. United States, 1976*. DHEW Publication No. (PHS) 80–1915, Washington, DC: US Government Printing Office.

Takahashi, E. (1984). Secular trend in milk consumption and growth in Japan. *Human Biology*, **56**, 427–37.

Tamarkin, L., Baird, C. J. & Almeida, O. F. X. (1985). Melatonin: a coordinating signal for mammalian reproduction? *Science*, **227**, 714–20.

Tanner, J. M. (1947). The morphological level of personality. *Proceedings of the Royal Society of Medicine*, **40**, 301–3.

Tanner, J. M. (1962). *Growth and Adolescence*, 2nd edn. Oxford: Blackwell Scientific Publications.

Tanner, J. M. (1963). The regulation of human growth. *Child Development*, **34**, 817–47.

Tanner, J. M. (1965). Radiographic studies of body composition in children and adults. In *Human Body Composition*, ed. J. Brozek, pp. 211–36. Oxford: Pergamon Press.

Tanner, J. M. (1969). Relation of body size, intelligence, test scores, and social circumstances. In *Trends and Issues in Developmental Psychology*, ed. P. Mussen, J. Langer & M. Covington, pp. 182–201. New York: Holt, Rinehart & Winston.

Tanner, J. M. (1978). *Fetus Into Man*. Cambridge: Harvard University Press.

Tanner, J. M. (1981). *A History of the Study of Human Growth*. Cambridge: Cambridge University Press.

Tanner, J. M. & Cameron, N. (1980). Investigation of the mid-growth spurt in height, weight and limb circumferences in single-year velocity data from the London 1966–67 growth survey. *Annals of Human Biology*, **7**, 565–77.

Tanner, J. M. & Eveleth, P. B. (1976). Urbanization and growth. In *Man in*

Urban Environments, ed. G. A. Harrison, J. B. Gibson, pp. 144–66. Oxford: Oxford University Press.

Tanner, J. M., Prader, A., Habich, H. & Ferguson-Smith, M. A. (1959). Genes on the Y chromosome influencing rate of maturation in man: skeletal age studies in children with Klinefelter's (XXY) and Turner's (XO) syndromes. *Lancet*, **2**, 141–4.

Tanner, J. M., Whitehouse, R. H., Hughes, P. C. R. & Carter, B. S. (1976a). Relative importance of growth hormone and sex steroids for the growth at puberty of trunk length, limb length, and muscle width in growth hormone deficient children. *Journal of Pediatrics*, **89**, 1000–8.

Tanner, J. M., Whitehouse, R. H., Marubiri, E. & Resele, F. (1976b). The adolescent growth spurt of boys and girls of the Harpenden Growth Study. *Annals of Human Biology*, **3**, 109–26.

Tanner, J. M., Hayashi, T., Preece, M. A. & Cameron, N. (1982). Increase in length of leg relative to trunk in Japanese children and adults from 1957 to 1977: comparison with British and with Japanese Americans. *Annals of Human Biology*, **9**, 411–23.

Tanner, J. M., Landt, K. W., Cameron, N., Carter, B. S. & Patel, J. (1983a). Predicting adult height from height and bone age in childhood. *Archives of Disease in Childhood*, **58**, 767.

Tanner, J. M., Whitehouse, R. H., Cameron, N., Marshall, W. A., Healy, M. J. R. & Goldstein, H. (1983b). *Assessment of Skeletal Maturity and Prediction of Adult Height*, 2nd edn. London: Academic Press.

Tanner, N. & Zihlman, A. (1976). Women in evolution. Part I: Innovation and selection in human origins. *Signs*, **1**, 585–608.

Taranger, J., Engstrom, I., Lichenstein, H. & Svennberg-Redegren, I. (1976). Somatic pubertal development. *Acta Paediatrica Scandinavica*, Suppl., **258**, 121–35.

Tattersal, I. (1975). *The Evolutionary Significance of Ramapithecus*. Minneapolis: Burgess.

Teleki, G. E., Hunt, E. & Pfifferling, J. H. (1976). Demographic observations (1963–1973) on the chimpanzees of the Gombe National Park, Tanzania. *Journal of Human Evolution*, **5**, 559–98.

Tetsuo, M., Poth, M. & Markey, S. P. (1982). Melatonin metabolite excretion during childhood and puberty. *Journal of Clinical Endocrinology and Metabolism*, **55**, 311–13.

Thom, R. (1975) *Structural stability and morphogenesis*. Translated by D. E. Fowler. Reading: Benjamin.

Thom, R. (1983). *Mathematical models of morphogenesis*. Translated by W. M. Brookes & D. Rand. New York: Halsted Press/John Wiley.

Thompson, D'Arcy, W. (1917). *On Growth and Form*. Cambridge: Cambridge University Press.

Thompson, D'Arcy, W. (1942). *On Growth and Form*, revised edition. Cambridge: Cambridge University Press.

Timiras, P. S. (1972). *Developmental Physiology and Aging*. New York: Macmillan Publishing Co.

Tisserand-Perier, M. (1953). Etudes de certains processus de croissance chez les jumeaux. *Journal de Genetic Humaine*, **2**, 87–102.

Tobias, P. V. (1975). Anthropometry among disadvantaged peoples: studies in

Southern Africa. In *Biosocial Interrelations in Population Adaptation*, ed. E. S. Watts, F. E. Johnston & G. W. Lasker, pp. 287–305. The Hague: Mouton.

Tobias, P. V. (1985). The negative secular trend. *Journal of Human Evolution*, **14**, 347–56

Tobias, P. V., Netscher, D. (1976). Evidence from African Negro skeletons for a reversal of the usual secular trend (abstract). *Journal of Anatomy*, **121**, 435–6.

Todd, J. T., Mark, L. S., Shaw, R. E. & Pittenger, J. B. (1980). The perception of human growth. *Scientific American*, **242**(2), 132–44.

Todd, T. W. (1937). *Atlas of Skeletal Maturation*. St Louis: C. V. Mosby.

Trivers, R. L. (1972). Parental investment and sexual selection. In *Sexual Selection and the Descent of Man*, ed. B. Campbell, pp. 136–79. Chicago: Aldine.

Turnbull, C. M. (1983a). *The Human Cycle*, pp. 43–4. New York: Simon & Schuster.

Turnbull, C. M. (1983b). *The Mbuti Pygmies*. New York: Holt, Rinehart & Winston.

United Nations, (1980). *Patterns of Urban and Rural Growth*. Population Studies, No. 68. New York: United Nations.

Van de Hulst, H. C. (1957). *Light Scattering by Small Particles*. New York: Wiley.

VanLoon, H., Saverys, V., Vuylsteke, J. P., Vlietinck, R. F. & Eeckels, R. (1986). Local versus universal growth standards: the effect of using NCHS as a universal reference. *Annals of Human Biology*, **13**, 347–57.

Varrela, J. (1984). Effects of X chromosome on size and shape of body: an anthropometric investigation in 47,XXY males. *American Journal of Physical Anthropology*, **64**, 233–42.

Varrela, J. & Alvesalo, L. (1985). Effects of the Y chromosome on quantitative growth: An anthropometric study of 47,XYY males. *American Journal of Physical Anthropology*, **68**, 239–45.

Varrela, J., Alvesalo, L. & Vinkka, H. (1984a). The phenotype of 45,X females: an anthropometric quantification. *Annals of Human Biology*, **11**, 53–66.

Varrela, J., Alvesalo, L. & Vinkka, H. (1984b). Body size and shape in 46,XY females with complete testicular feminization. *Annals of Human Biology*, **4**, 291–301.

Vaughn, J. M. (1975). *The Physiology of Bone*, 2nd edn. Oxford: Clarendon Press.

Vavra, H. M. & Querec, L. J. (1973). *A Study of infant mortality from linked records by age of mother, total-birth order, and other variables*. DHEW Publication No. (HRA) 74–1851. Washington, DC: US Government Printing Office.

Vetta, A. (1975). Fertility, physique, and intensity of selection. *Human Biology*, **47**, 283–93.

Villar, J. & Belizan, J. M. (1982). The relative contribution of prematurity and fetal growth retardation to low birth weight in developing and developed countries. *American Journal of Obstetrics and Gynecology*, **143**, 793A–8A.

Vincent, M. & Dierickx, J. (1960). Etude sur la croissance saisonnaire des escoliers de Leopoldville. *Annales de la Societé Belge de Médecine Tropicale*, **40**, 837–44.

Waldhauser, F., Weizenbacher, G., Frisch, H., Zeitalhuber, U., Waldhauser, N.

& Wurtman, R. J. (1984). Fall in nocturnal serum melatonin during pre-puberty and pubescence. *Lancet*, **1**, 362–5.

Walker, A. C. (1981). Dietary hypotheses and human evolution. *Philosophical Transactions of the Royal Society*, B**292**, 58–64.

Walker, E. P. (1975). *Mammals of the World*, 3rd edn. Baltimore: Johns Hopkins University Press.

Warren, M. P. (1980). The effects of exercise on pubertal progression and reproductive function in girls. *Journal of Clinical Endocrinology and Metabolism*, **51**, 1150–7.

Washburn, S. L. (1981). Longevity in primates. In *Aging: Biology and Behavior*, ed. J. L. McGaugh & S. B. Kiesler, pp. 11–29. New York: Academic Press.

Waterlow, J. C. (1975). Protein turnover in the whole body *Nature*, **253**, 157.

Waterlow, J. C. & Payne, P. R. (1975). The protein gap. *Nature*, **258**, 113–17.

Waterlow, J. C., Buzina, R., Keller, W., Lane, J. M., Nichaman, M. Z. & Tanner, J. M. (1977). The presentation and use of height and weight data for comparing the nutritional status of children under the age of 10 years. *Bulletin of the World Health Organization*, **55**, 489–98.

Watts, E. S. (1985). Adolescent growth and development of monkeys, apes and humans. In *Nonhuman Primate Models for Human Growth and Development*, ed. E. S. Watts, pp. 41–65. New York: Alan R. Liss.

Watts, E. S. (1986). The evolution of the human growth curve. In *Human Growth*, Vol. 1, 2nd edn, ed. F. Faulkner & J. M. Tanner, pp. 153–6. New York: Plenum.

Watts, E. S. & Gavan, J. A. (1982). Postnatal growth of nonhuman primates: the problem of adolescent spurt. *Human Biology*, **54**, 53–70.

Weirman, M. E. & Crowley, W. F. Jr. (1986). Neuroendocrine control of the onset of puberty. In *Human Growth*, Vol. 2, 2nd edn, ed. F. Falkner & J. M. Tanner, pp. 225–41. New York: Plenum.

Weiss, P. & Kavanau, J. L. (1957). A model of growth and growth control in mathematical terms. *Journal of General Physiology*, **41**, 1–47.

Weitzman, E. B., Boyar, R. M., Kapen, S. & Hellman, L. (1975). The relation-ship of sleep and sleep stages to neuroendocrine secretion and biological rhythms in man. *Recent Progress in Hormone Research*, **41**, 339–446.

Werner, E. E., Bierman, J. M. & French, F. E. (1971). *The Children of Kauai*. Honolulu: University of Hawaii Press.

Widdowson, E. M. (1951). Mental contentment and physical growth. *Lancet*, **1**, 1316–18.

Widdowson, E. M. (1968). The place of experimental animals in the study of human malnutrition. In *Caloric Deficiencies and Protein Deficiencies*, ed. R. A. McCance & E. M. Widdowson, pp. 225–36. London: J. & A. Churchill.

Widdowson, E. M. (1970). The harmony of growth. *Lancet*, **1**, 901–5.

Widdowson, E. M. (1976). Pregnancy and lactation: the comparative point of view. In *Early Nutrition and Later Development*, ed. A. W. Wilkinson, pp. 1–10. Chicago: Year Book Medical Publishers.

Widdowson, E. M. & Dickerson, J. W. T. (1964). The chemical composition of the body. In *Mineral Metabolism*, Vol. 2A, ed. C. L. Comar & F. Bronner, pp. 1–247: New York: Plenum.

Wilson, R. S. (1979). Twin growth: initial deficit, recovery, and trends in concordance from birth to nine years. *Annals of Human Biology*, **6**, 205–20.

Winter, J. S. D. (1978). Prepubertal and pubertal endocrinology. In *Human Growth*, Vol. 2, ed. F. Falkner & J. M. Tanner, pp. 183–213. New York: Plenum.

Wolff, G. (1935). Increased bodily growth of school-children since the war. *Lancet*, 1, 1006–11.

Wolpoff, M. H. (1980). *Paleoanthropology*. New York: Knoff.

Woodcock, A. & Davis, M. (1978). *Catastrophe Theory*. New York: Dutton.

Worthman, C. M. (1986). Later-maturing populations and control of the onset of puberty. *American Journal of Physical Anthropology*, **69**, 282 (abstract).

Wurtman, R. J. (1975). The effects of light on the human body. *Scientific American*, **233**(1), 68–77.

Wurtman, R. J. & Axelrod, J. (1965). The pineal gland. *Scientific American*, **223**, 50–60.

Young, V. R., Steffee, W. P., Pencharz, P. B., Winterer, J. C. & Scrimshaw, N. S. (1975). Total human body protein synthesis in relation to protein requirements at various ages. *Nature*, **253**, 192–4.

Zeeman, E. C. (1976). Catastrophe theory. *Scientific American*, **234**, 65–83.

Zeeman, E. C. (1977). *Catastrophe Theory: selected papers 1972–1977*. Reading: Benjamin.

Zemel, B. S. & Katz, S. H. (1986). The contribution of adrenal and gonadal androgens to the growth in height of adolescent males. *American Journal of Physical Anthropology*, **71**, 459–66.

Zezulak, K. M. & Green, H. (1986). The generation of insulin-like growth factor-1-sensitive cells by growth hormone action. *Science*, **233**, 551–3.

Index

Page numbers in italic type refer to figures, those in bold to tables.